T0192614

APPLIED GEOSTATISTICS WITH SGeMS

A User's Guide

The Stanford Geostatistical Modeling Software (SGeMS) is an open-source computer package for solving problems involving spatially related variables. It provides geostatistics practitioners with a user-friendly interface, an interactive 3D visualization and a wide selection of algorithms. With over 12,000 downloads in less than 2 years, SGeMS is used in several research groups and companies.

This practical book provides a step-by-step guide to using SGeMS algorithms. It explains the underlying theory, demonstrates the implementation of the various algorithms, discusses their potential limitations, and helps the user make an informed decision about the choice of one algorithm over another. Users can complete complex tasks using the embedded scripting language, and new algorithms can be developed and integrated through the SGeMS plug-in mechanism. SGeMS is the first software to provide algorithms for multiple-point statistics and the book presents an up-to-date discussion of the corresponding theory and applications.

Incorporating the full SGeMS software, this book is an essential userguide for professional practitioners of environmental, mining and petroleum engineering, as well as graduate students and researchers in fields such as remote sensing, geography, ecology, water resources and hydrogeology. Both beginners and more advanced users will find answers in the book on how to use the software and more generally about geostatistical practice. Navigate to www.cambridge.org/9781107403246 for the SGeMS software.

NICOLAS REMY received a B.S. in Mathematics and Physics from École Nationale Supérieure des Mines, Nancy, France, an M.S. in Petroleum Engineering from Stanford University and a Ph.D. in Geostatistics from Stanford University. He is currently a Senior Statistician at Yahoo!, leading the Data Mining and User Behavior Modeling group for the Yahoo! Media and Yahoo! Communications and Communities business units. His research interests include multiple-points statistics, machine learning, graph theory and data mining.

ALEXANDRE BOUCHER received a B.Eng. in Geological Engineering from the École Polytechnique de Montréal, Montreal, QC, Canada, an M.Phil. degree from the University of Queensland, Brisbane, Australia, and a Ph.D. from Stanford University, Stanford, CA. He teaches geostatistics in the Department of Environmental Earth System Science, Stanford University, and has taught short courses on the subject in the USA and Japan. His research interests include geostatistics, data integration, remote sensing, uncertainty modeling, machine learning, and probabilistic modeling of spatio-temporal phenomena.

JIANBING WU received his Ph.D. in Petroleum Engineering in 2007 from Stanford University, and his M.E. and B.S. degrees in Mechanical Engineering from University of Science and Technology of China. He is a reservoir engineer with the Applied Reservoir Engineering group at ConocoPhillips, and his research focuses on static and dynamic reservoir modeling. He is currently a member of SPE, IAMG and SEG.

APPLIED GEOSTATISTICS
WITH SGeMS
A User's Guide

NICOLAS REMY

Yahoo! Media and Yahoo! Communications and Communities

ALEXANDRE BOUCHER

Stanford University

and

JIANBING WU

ConocoPhillips

CAMBRIDGE UNIVERSITY PRESS
Cambridge, New York, Melbourne, Madrid, Cape Town,
Singapore, São Paulo, Delhi, Tokyo, Mexico City

Cambridge University Press
The Edinburgh Building, Cambridge CB2 8RU, UK

Published in the United States of America by Cambridge University Press, New York

www.cambridge.org
Information on this title: www.cambridge.org/9781107403246

© N. Remy, A. Boucher and J. Wu 2009

First published 2009
First paperback edition 2011

A catalogue record for this publication is available from the British Library

Additional resources for this publication at www.cambridge.org/9781107403246

ISBN 978-0-521-51414-9 Hardback
ISBN 978-1-107-40324-6 Paperback

Contents

v

Foreword

Geostatistics is a science. It is also an art.

Geostatistics is not simply the application of statistical methods to geology-driven spatial distributions, it also provides a conceptual framework for making inferences from Earth sciences data – data which are, more often than not, incomplete.

Some may say, as I would, that most of the problems addressed in geostatistics are inverse problems, in the sense that data are used to infer parameters of the prior model. However, a gap exists between inverse problems and geostatistical problems: in inverse problems modeling the observations can be computer intensive, but the a priori model is typically simple (or simplistic); in a geostatistical problem the data are directly related to the model parameters, this allows one to consider prior models that encapsulate properties of the real Earth, sometimes with breathtaking realism. The gap is narrowing and will disappear in the near future. For the time being, we should try to do the best we can in each of the two fields.

Most geostatistical solutions involve random functions, and a long way has been trod since the simple Gaussian models used in the beginning. The Stanford team has been at the origin of many non-Gaussian developments that have now become standard. They are at it again with the multiple-point geostatistical concept and related algorithms, which allow defining truly complex random functions.

If a painter has no brushes, or no paints, she/he can not produce art. The same happens here: in order to deal with difficult problems calling for non-simplistic priors and working out practical solutions (the art), one needs computer software. The equations underwriting geostatistics can be laid on paper, but even the simplest application requires dedicated computer software. This is where the Stanford Geostatistical Modeling Software (SGeMS) shines. With this book the reader will learn how to use the software towards solving non-trivial problems.

Nothing can replace my repeated stays at Stanford and interaction with the people there at the Center for Reservoir Forecasting – with, among others, the authors of this book and my dear friend André Journel. Having this book with me is the closest thing to being at Stanford.

Albert Tarantola
Pasadena

Preface

This book is not an introduction to geostatistics and its theory. Although some elements of geostatistics theory are recalled, the text assumes a reasonable level of familiarity with the main concepts of geostatistics: notions of a random function, stationarity or variogram should not leave you wondering.

The main purpose of this book is to back up the Stanford Geostatistical Modeling Software (SGeMS) and, hopefully, widen the reader's comprehension of geostatistics beyond its theory into its very diverse applications. In that perspective, the emphasis is on practical aspects (in what context should one tool be preferred over another one) and implementation considerations (to what extent do the algorithm implementations deviate from the theory, what are the assumptions and limitations of the implementation). However, this book is not a reference manual to programming in SGeMS and no details about the source code of SGeMS or its APIs are given. If you are interested in learning how to contribute code to SGeMS please refer to the SGeMS web site, http://sgems.sourceforge.net, where a description of the SGeMS APIs and several tutorials are available.

The geostatistics tools of SGeMS were designed to provide enough flexibility to adapt to very different problems. As a result, the number of available control parameters may seem daunting to the novice practitioner. Don't be intimidated! Most of the advanced parameters have default values, and the best way to build up your understanding of those parameters is to repeat the example runs and experiment on your own.

While most of the tools in SGeMS are based on classical geostatistics (kriging, Gaussian simulation, indicator simulation, etc.), a large portion of the book is devoted to the concept of multiple-points statistics. Multiple-points statistics being a new and promising area of geostatistics, its underlying theory is exposed in greater detail, and two multiple-point algorithms are thoroughly described.

This book has had a long gestation. The idea of a generic geostatistics programming library (the GsTL library), accompanied by a simple showcase software dates back to 2001 and was started with the collaboration of Professor Arben Schtuka, then at École Nationale Supérieure de Géologie (France), and the support of Professor Jef Caers at Stanford University. What was supposed to be a simple software, however, turned into a very involved programming effort. SGeMS is now a full fledged software that provides a modern, convenient and powerful platform for new developments of geostatistics. Thanks to the support of Jef Caers and André Journel, the SGeMS software gained traction at the Stanford Center for Reservoir Forecasting (SCRF), then with its industrial affiliates and collaborating universities.

This book wouldn't have existed without the support of André Journel, his untiring cheer-leading, dedicated supervision and his obsessive proof-reading. We are also very grateful to Dr. Mohan Srivastava, Dr. Ricardo Olea and Dr. Pierre Goovaerts who carefully reviewed the early drafts of the manuscript and meticulously tracked the inconsistencies and many bugs of the software. The text and software greatly benefited from their many suggestions and remarks. We would like to thank Dr. Sébastien Strebelle, Professor Sanjay Srinivasan and Professor Guillaume Caumon, who carefully reviewed the final versions of the manuscript. Finally, we are very grateful to Professor Jef Caers for initiating and strongly believing in the GsTL project: without his support, SGeMS wouldn't have been written.

By design, SGeMS is not a static and complete software. New algorithms will be added, and its underlying API may change. We welcome comments on it, reports of bugs, valuable enhancement ideas or source code. You can send comments to the SGeMS mailing-lists and refer to the SGeMS web site at `http://sgems.sourceforge.net` for updates and code documentation.

List of programs

Exploratory data analysis

Histogram: histogram plot and statistics, 85
QQ/pageref-plot: Q-Q or P-P plot and statistics, 87
Scatter-plot: scattergram plot and statistics, 87
Variogram: variogram computation and modeling, 90

Estimation

BKRIG: block kriging, 122
COKRIGING: kriging with secondary data, 59
INDICATOR KRIGING: indicator kriging, 113
KRIGING: kriging, 109

Simulation

BESIM: block error simulation, 163
BSSIM: block sequential simulation, 157
COSGSIM: sequential Gaussian co-simulation, 139
COSISIM: sequential indicator co-simulation, 153
DSSIM: direct sequential simulation, 143
FILTERSIM: filter-based simulation, 191
LUSIM: LU simulation, 133
SGSIM: sequential Gaussian simulation, 135
SISIM: sequential indicator simulation, 147
SNESIM: single normal equation simulation, 169

Utility

List of symbols

cdf	Cumulative distribution function
E-type	Conditional expectation estimate obtained by point-wise averaging of simulated realizations
EDA	Elementary data analysis
FFT	Fast Fourier Transform
GSLIB	Geostatistical software library, as in Deutsch and Journel (1998)
IK	Indicator kriging
KT	Kriging with a trend
LVM	Kriging with a local varying mean
M-type	Conditional median estimate
MM1	Markov Model 1
MM2	Markov Model 2
mp	Multiple-point
OK	Ordinary kriging
P-P plot	Probability–probability plot
pdf	Probability density function
Q-Q plot	Quantile–quantile plot
RF	Random function
RV	Random variable
SGeMS	Stanford Geostatistical Modeling Software
SK	Simple kriging
Ti	Training image
$\perp$	Orthogonal
$\forall$	Whatever

α, β, γ	Rotation angles for -azimuth, -dip and rake
$\gamma(\mathbf{h})$	Stationary semivariogram model
$\gamma(Z(\mathbf{u}_\alpha), Z(\mathbf{u}_\beta))$	Semivariogram value between any two random variables $Z(\mathbf{u}_\alpha)$ and $Z(\mathbf{u}_\beta)$
$\gamma^*(\mathbf{h})$	Experimental semivariogram
$\gamma^{(l)}$	lth component of a nested semivariogram model
$\gamma_{ij}(\mathbf{h})$	Cross semi-variogram model between any two random variables $Z_i(\mathbf{u})$ and $Z_j(\mathbf{u} + \mathbf{h})$
$\lambda_\alpha, \lambda_\alpha(\mathbf{u})$	Kriging weight associated to datum location $\mathbf{u}_\alpha$ for estimation at location $\mathbf{u}$. The superscripts (SK), (OK), (KT) are used when necessary to differentiate between various types of kriging
Γ	Kriging weights matrix
Λ	Scaling matrix
λ	Column matrix of the $n(\mathbf{u})$ kriging weights
Θ	Rotation matrix
ν_i	nu parameter for the ith attribute
ω	Parameter of a power function; or a servosystem factor
$\phi_{\text{lti}}(\cdot)$	Low tail extrapolation function
$\phi_{\text{uti}}(\cdot)$	Upper tail extrapolation function
$\rho(\mathbf{h})$	Stationary correlogram $\in [-1, +1]$
σ^2	Variance
$\sigma^2_{\text{SK}}(\mathbf{u})$	Kriging variance of $Z(\mathbf{u})$. The subscripts (SK), (OK), (KT) are used when necessary to differentiate between various types of kriging
T_J^g	Expanded search template in the gth multiple grid
τ_i	tau parameter for the ith attribute
T_J	A search template with J nodes
$\bar{C}_{BB'}, \bar{C}(V, V')$	Block-to-block covariance model
$\bar{C}_{PB}, \bar{C}(\mathbf{u}, V(\mathbf{s}))$	Point-to-block covariance model
$\mathbf{h}$	Coordinates offset vector, or a lag vector
$\mathbf{h}_j$	Offset of the jth node in the search template from its center
$\mathbf{D}$	Column matrix of $n(\mathbf{u})$ residual data values $z(\mathbf{u}_\alpha) - m$
$\mathbf{D}_i$	Data vector involving i multiple data locations $\{D_i = d_i, i = 1, \ldots, n\}$
$\mathbf{K}$	Data-to-data square covariance matrix
$\mathbf{k}$	Data-to-unknown covariance matrix
$\mathbf{prot}$	Prototype of a categorical variable
$\tilde{A}$	No-A

$\mathbf{u}$	Coordinates vector
$\mathbf{u}_\alpha, \mathbf{u}_\beta$	Data locations
$\mathbf{v}$	Block volume, or a set of points
a	Range parameter
$a_i(\mathbf{u})$	Coefficient of component number k of the trend model
B	Block data
$B(\mathbf{v}_\alpha)$	A linear average value within a block volume $\mathbf{v}_\alpha$
$B_V(\mathbf{s})$	A linear average value within a block V centered at location $\mathbf{s}$
$C(0)$	Covariance value at separation vector $\mathbf{h} = 0$. It is also the stationary variance of random variable $Z(\mathbf{u})$
$C(\mathbf{h})$	Covariance between any two random variables $Z(\mathbf{u})$ and $Z(\mathbf{u} + \mathbf{h})$ separated by vector $\mathbf{h}$
c_l	Variance contribution of the lth nested semi-variogram model
C_R	Error covariance matrix
$C_{ij}(\mathbf{h})$	Cross covariance between any two random variables $Z_i(\mathbf{u})$ and $Z_j(\mathbf{u})$ separated by vector $\mathbf{h}$
$cmin$	Minimum number of pattern replicates
dev	Local conditioning data event
dev_J	Local conditioning data event found by search template T_J
$E\{\cdot\}$	Expected value
$\text{Exp}(\cdot)$	Exponential semi-variogram function
F	Number of filters
$f(\mathbf{h}_j)$	Filter weight associated with the jth template node
$F(\mathbf{u}, z)$	Cumulative distribution function of random variable $Z(\mathbf{u})$
$F(z)$	Cumulative histogram of RV
$f(z)$	Probability density function or histogram
$F^{-1}(p)$	Inverse cumulative distribution function or quantile function for the probability value $p \in [0, 1]$
f_x, f_y, f_z	Affinity factor in each X/Y/Z direction
F_Z	Marginal cdf of random function Z
$G(\cdot)$	Standard normal cumulative distribution function
$G^{-1}(p)$	Standard normal quantile function such that $G(G^{-1}(p)) = p \in [0, 1]$
h_x, h_y, h_z	Variogram ranges in X/Y/Z directions
$I(\mathbf{u}; z_k)$	Binary indicator random function at location $\mathbf{u}$ and for cutoff z_k
$i(\mathbf{u}; z_k)$	Binary indicator value at location $\mathbf{u}$ and for cutoff z_k
$I^*(\mathbf{u}; z_k)$	Indicator estimator for cutoff z_k
$i^*(\mathbf{u}; z_k)$	Indicator kriging estimated value for cutoff z_k

$I_{SK}^*(\mathbf{u})$	Indicator kriging estimator of categorical indicator $I_k(\mathbf{u})$
$I_k(\mathbf{u})$	Binary indicator random function at location $\mathbf{u}$ and for category k
$i_k(\mathbf{u})$	Binary indicator value at location $\mathbf{u}$ and for category k
K	Number of categories
L_α	A known linear averaging function
M	Median of an RF
m	Mean value of a random variable
$m(\mathbf{u})$	Mean function at location $\mathbf{u}$; expected value of random variable $Z(\mathbf{u})$; or trend component model in the decomposition $Z(\mathbf{u}) = m(\mathbf{u}) + R(\mathbf{u})$, where $R(\mathbf{u})$ is the residual component model
$m^*(\mathbf{u})$	Estimate of the trend component or locally varying mean at location $\mathbf{u}$
$N(\mathbf{h})$	Number of data pairs separated by vector $\mathbf{h}$
$n(\mathbf{u})$	n conditioning data found in a neighborhood centered at $\mathbf{u}$
n_k	Number of patterns whose center takes a specific value k
P	Point data
p	Probability value
p_k^c	Current proportion of category k simulated so far
p_k^t	Target proportion of category k
p_0	Prior probability of the event occurring
pat	Training pattern
$\text{Prob}\{\cdot\}$	Probability function
$prot$	Prototype of a continuous variable
$q(p) = F^{-1}(p)$	Quantile function for the probability value $p \in [0, 1]$
$R(\mathbf{u})$	Residual random function model at location $\mathbf{u}$ in the decomposition $Z(\mathbf{u}) = m(\mathbf{u}) + R(\mathbf{u})$, where $m(\mathbf{u})$ is the trend component model
$r(\mathbf{u})$	Residual value at location $\mathbf{u}$
r^i	Azimuth rotation angle in rotation region i
$r_s(\mathbf{u})$	Simulated residual value at location $\mathbf{u}$
S	A set of locations $\mathbf{u}$, or a deposit
$S_T^k(\mathbf{u})$	The kth+ filter score value of pattern $pat(\mathbf{u})$ found by search template T
$Sph(\cdot)$	Spherical semi-variogram function
$t(\mathbf{u})$	Training image nodal value at location $\mathbf{u}$
$V, V(\mathbf{u})$	A block centered at location $\mathbf{u}$
$\text{Var}\{\cdot\}$	Variance
x_i	Prior distance to a given probability, used in the tau model

$Z(\mathbf{u})$	Generic random variable at location $\mathbf{u}$, or a generic random function of location $\mathbf{u}$
$z(\mathbf{u})$	Generic variable function of location $\mathbf{u}$
$z(\mathbf{u}_\alpha)$	z datum value at location $\mathbf{u}_\alpha$
$Z^*(\mathbf{u})$	Kriging estimator of $Z(\mathbf{u})$. The subscripts (SK), (OK), (KT) are used when necessary to differentiate between various types of kriging
$z^*(\mathbf{u})$	An estimate of value $z(\mathbf{u})$
$z_E^*(\mathbf{u})$	Conditional expectation, or E-type, obtained as point-wise arithmetic average of multiple realizations $z^{(l)}(\mathbf{u})$
$z_M^*(\mathbf{u})$	M-type estimated value, where $z_M^*(\mathbf{u})$ has a 50% chance to be higher (or lower) than the actual unknown value
$z_{K_s}^*(\mathbf{u})$	Kriging estimate built from the simulated value $z_s(\mathbf{u}_\alpha)$
$z_{LVM}^*(\mathbf{u})$	Estimated value with local varying mean at location $\mathbf{u}$
$z^{(l)}(\mathbf{u})$	lth realization of the random function $Z(\mathbf{u})$
$z_V^{(l)}(\mathbf{u})$	The simulated value over a block V averaged from the lth point support realization $z^{(l)}(\mathbf{u})$
z_k	kth threshold value for the continuous attribute z
$Z_{cs}(\mathbf{u})$	Conditional simulated random variable at location $\mathbf{u}$
$z_{cs}(\mathbf{u})$	Conditional simulated value at location $\mathbf{u}$

1

Introduction

SGeMS, the Stanford Geostatistical Modeling Software, is a software developed at Stanford University that implements several geostatistics algorithms for the modeling of earth systems and more generally space-time distributed phenomena. It was written with two goals in mind. The first one, geared toward the end-user, is to provide a user-friendly software which offers a large range of geostatistics tools: the most common geostatistics algorithms are implemented, in addition to more recent developments such as multiple-point statistics simulation. The user-friendliness of SGeMS comes from its non-obtrusive graphical user interface, and the possibility to directly visualize data sets and results in a full 3D interactive environment.

The second objective was to design a software that would cater to the needs of power-users. In SGeMS, most of the operations performed using the graphical interface can also be executed programmatically. The integrated support for the Python scripting language enables the creation of simple macros all the way to complete nested applications with an independent graphical interface. New features can conveniently be added into SGeMS through a system of plug-ins, i.e. pieces of software which can not be run by themselves but complement a main software. In SGeMS, plug-ins can be used to add new geostatistics tools, add new grid data structures (faulted stratigraphic grids for example) or define new import/export file filters. SGeMS is used as a development platform for geostatistics at the Stanford Center for Reservoir Forecasting (SCRF).

The book structure

Chapter 2 begins with a concise tutorial that walks the reader through the main steps involved in a simple geostatistical study. The aim of this tutorial is to give an overview of the capabilities of SGeMS and direct the reader to the different parts of the book for more details. The second part of Chapter 2 gives the conventions used throughout the book, for example, how 3D ellipsoids are defined, how Cartesian grids are defined and the details of data file formats.

Chapter 3 recalls the fundamental geostatistics concepts used in the book. Apart from classical aspects of geostatistics such as variograms and kriging, this chapter also introduces the concept of multiple-points statistics, at the root of two major algorithms presented in Section 8.2.

Chapter 4 presents the main data sets used throughout the rest of the book. All these data sets are available on the CD included with this book. As the data sets are described, the tools for elementary data exploration are introduced: histograms, scatterplots, quantile–quantile and probability–probability plots. Variograms being of particular importance in geostatistics are described in a separate chapter: Chapter 5 details the tools to compute experimental variograms and model them.

Chapters 6 through 9 constitute a reference manual to the SGeMS geostatistics algorithms. For each algorithm, practical aspects and implementation considerations are reviewed, the required input parameters are exhaustively described, and a brief example run is presented. The reader is encouraged to try to reproduce the runs to get familiar with the algorithm and its parameters. Chapter 6 introduces the main graphical interfaces used to input parameters to the SGeMS algorithms. For example, many algorithms such as kriging, sequential Gaussian simulation and direct sequential simulation require a variogram and a search ellipsoid. The variogram and search ellipsoid input interfaces used by these three algorithms are described in this chapter.

Chapter 7 describes the estimation algorithms available in SGeMS: simple kriging, ordinary kriging, kriging with a trend or a locally varying mean, indicator kriging and cokriging. The theory behind each algorithm is briefly recalled and implementation considerations are discussed. All the controlling parameters are then thoroughly described. Practical considerations are isolated in gray-background boxes.

Chapter 8 is divided into two main parts. Section 8.1 is dedicated to variogram-based simulation algorithms: sequential Gaussian simulation, sequential indicator simulation, direct sequential simulation and their co-simulation variants. The second half of Chapter 8 (Section 8.2) introduces two recently developed simulation algorithms: SNESIM and FILTERSIM. These two algorithms are based on the multiple-points statistics paradigm presented in Chapter 3. Because these algorithms are recent, a large portion of the text is devoted to the description of best practices and the impact of the input parameters on runtime performance and quality of the final results.

Chapter 9 presents service algorithms, called utilities, useful to prepare the input data of the estimation and simulation algorithms, and then analyze their output.

The last chapter (Chapter 10) teaches the advanced user how to automate tasks in SGeMS, using either its system of *commands* or the embedded Python scripting language. Describing Python would be a book topic of its own; this chapter just describes how SGeMS and Python cooperate, and directs the reader who wants to learn Python to outside sources. Last, a major feature of SGeMS is introduced: SGeMS supports a *plug-in* mechanism to augment its function-alities, allowing for example the addition of new geostatistics algorithms or adding support for new types of grids on which geostatistics could be performed. As for Python, a tutorial on C++ and SGeMS plug-ins development is beyond the scope of this book, and the advanced user is pointed to several on-line resources.

Source code and compilation

SGeMS is currently available on both Linux and Microsoft Windows platforms. Although it has also been successfully compiled on other Unix platforms (BSD and Solaris) and Mac OSX, no binaries are currently available for those operating systems. The code is distributed under the GNU General Public License (GPL). For more information about the GPL, refer to `http://www.gnu.org/copyleft/gpl.html`.

The source code and a Microsoft Windows executable are distributed on the accompanying disc, and can also be downloaded from the web at `http://sgems.sourceforge.net`.

To compile SGeMS, the following third-party libraries are required.

- GsTL (Geostatistics Template Library)
- Qt (GUI library) version 3.x (version 4 and higher is not supported yet)
- Coin3D (OpenInventor library), version 2.x
- SoQt (Qt bindings for OpenInventor), version 1.x
- SimVoleon (Volume rendering extension for Coin3D), version 2.x

A compiler that correctly supports C++ templates (e.g. member templates and template specialization) is also required. SGeMS has been successfully compiled with gcc-2.96, gcc-3.3.4, gcc-4, Intel C++ compiler, Visual C++ 2003 and Visual C++ 2005.

2

General overview

2.1 A quick tour of the graphical user interface

The graphical user interface (GUI) of SGeMS is divided into three main parts, see Fig. 2.1.

The Algorithm Panel The user selects in this panel which geostatistics tool to use and inputs the required parameters (Fig. 2.2). The top part of that panel shows a list of available algorithms, e.g. kriging, sequential Gaussian simulation. When an algorithm from that list is selected, a form containing the corresponding input parameters appears below the tools list.

The Visualization Panel This panel is an interactive 3D environment in which one or multiple objects, for example a Cartesian grid and a set of points, can be displayed. Visualization options such as color-maps are also set in the Visualization Panel. The Visualization Panel is shown in more detail in Fig. 2.3.

The Command Panel This panel gives the possibility to control the software from a command line rather than from the GUI. It displays a history of all previously executed commands and provides an input field where new commands can be typed (Fig. 2.4). See tutorial Section 2.2 and Chapter 10 for more details about the Command Panel. Note that the Command Panel is not shown by default when SGeMS is started: select *Command Panel* from the *View* menu to display it.

2.2 A typical geostatistical analysis using SGeMS

This short tutorial gives an overview of the capabilities of SGeMS and can serve as a "getting started" guide. It is rather fast paced and some beginners may find it overwhelming. If that is the case, we suggest you skim through it and come back to it later. The tutorial describes a SGeMS session in which a variable, rock porosity, is estimated at several unsampled locations using the "simple kriging" algorithm.

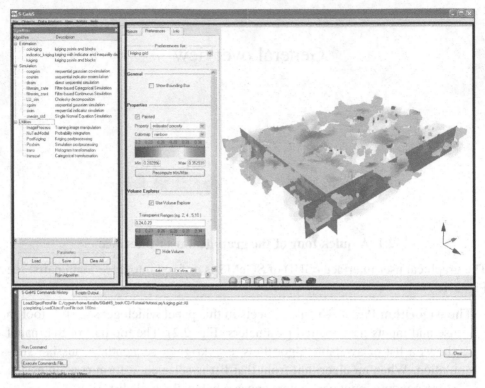

Figure 2.1 SGeMS's graphical interface. The three main panels are highlighted.
The top left panel is the Algorithm Panel, top right is the Visualization Panel and
bottom panel is the Command Panel

The data set used in this tutorial relates to a (synthetic) oil reservoir that displays
sand channels aligned in the North–South direction, with a proportion of sand (net-
to-gross ratio) of 33%. The reservoir is discretized into $100 \times 130 \times 10$ identical
parallelepipedic blocks (*cells*) in the X, Y and Z directions respectively. In the rest
of the book, this type of discretization is referred to as a Cartesian grid. The vari-
ations of rock porosity within a cell are considered negligible, and the problem of
estimating rock porosity in each block is simplified to estimating porosity at the
center of the cells (*cell-centered grid*).

Figure 2.5 shows a 3D view of the reference model with the sand channels in
black.

SGeMS provides several estimation tools, most of them based on the *kriging*
algorithm. In this tutorial *simple kriging* is used to estimate the porosity of the sand
channels. The reader unfamiliar with simple kriging is invited to refer to Section 3.6
for a brief presentation of kriging and references.

Here, 563 samples of porosity values were collected in the sand areas. The steps
involved in estimating porosity in the sand channels are as follows.

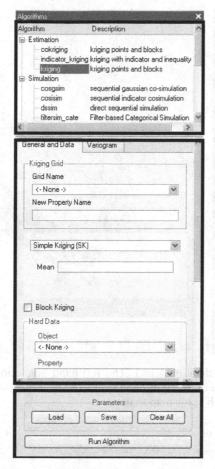

Figure 2.2 The three parts of the Algorithm Panel are highlighted. The top part displays the list of available algorithms. The middle section is where the input parameters for the selected algorithm are entered. The bottom part contains the controls to load/save the algorithms' parameters, and run the selected algorithm

1. Load the 563 samples data set in SGeMS.
2. Elementary data analysis: visualize the experimental porosity distribution and compute statistics such as quartiles, experimental mean and variance.
3. Compute the experimental variogram of porosity and model it.
4. Create the Cartesian grid on which kriging will be performed. Porosity is estimated at every node of that grid, whether or not the node belongs to a sand channel.
5. Select the *simple kriging* tool and enter the necessary parameters.
6. Display the results.
7. Post-process the results to retain only the region of interest (sand channels).
8. Save the results.

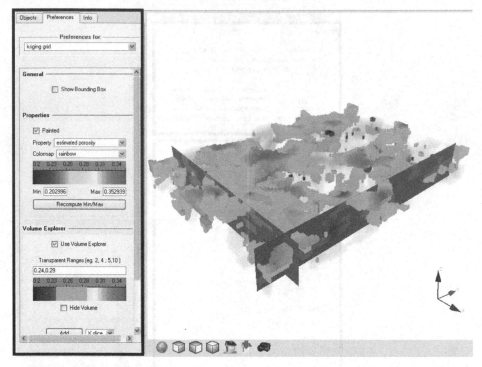

Figure 2.3 The Visualization Panel. The left-hand side controls which objects (e.g. grids) are visible in the right-hand side window. It is also used to set display options, such as which color-map to use

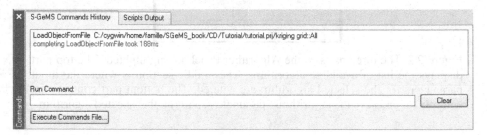

Figure 2.4 The Command Panel

2.2.1 *Loading data into an SGeMS project*

SGeMS calls *project* the set of all the objects currently loaded in its objects database. When SGeMS starts, it creates an empty project by default. Any object (Cartesian grid or set of points) later loaded into SGeMS is added to that project.

File sample_data.gslib on the CD contains the X, Y, Z coordinates and the porosity of each of the 563 random samples. It is an ASCII file, following the GSLIB format. Refer to Section 2.3 for a description of the GSLIB file format and other available data file formats.

Figure 2.5 3D view of the tutorial oil reservoir

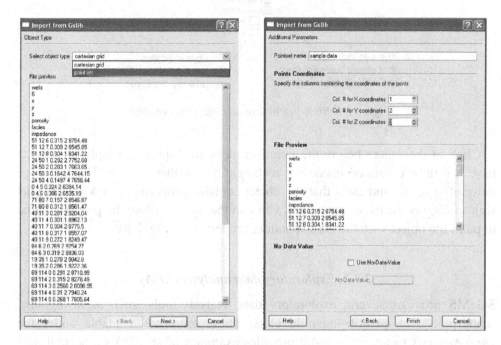

Figure 2.6 Load GSLIB point-set file wizard

To load the file, click *Objects* → *Load Object* and browse to the file location (or drag-and-drop the file onto the Visualization Panel). Since the GSLIB file format does not provide all the information required by SGeMS, a wizard prompting for the additional information pops up (see Fig. 2.6). The first screen of the wizard prompts for the type of the object, a set of points or a Cartesian grid, described by the file. Choose "Point set" and click **Next**. In the second screen, give a name for the point-set, e.g. `sample data`, and indicate which columns contain the X, Y and Z coordinates of the points (in this case, columns 1, 2 and 3).

Once the object is loaded a new entry called `sample data` appears in the **Objects** section of the Visualization Panel, as shown in Fig. 2.7.

Figure 2.7 Object list after the data set is loaded

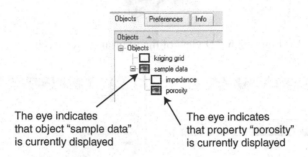

The eye indicates
that object "sample data"
is currently displayed

The eye indicates
that property "porosity"
is currently displayed

Figure 2.8 Showing/hiding an object or a property

Click in the square before the point-set name to display it. Displayed objects have a little eye painted inside the rectangle before their names. The plus sign before the square indicates that the object contains properties. Click on the plus sign to display the list of properties. Click in the square before the property name to paint the object with the corresponding property (see Fig. 2.8).

2.2.2 Exploratory data analysis (EDA)

SGeMS provides several exploratory data analysis tools, such as histograms, scatterplots and quantile–quantile plots. These tools are available under the *Data Analysis* menu. Section 4.2 provides examples of an EDA for several data sets, with more details on the data analysis tools of SGeMS.

The histogram of the sample porosity data is shown in Fig. 2.9.

2.2.3 Variogram modeling

Estimating porosity with simple kriging requires knowledge of the mean and the semi-variogram of the porosity variable. This semi-variogram can be modeled from the experimental variogram computed from the 563 sample points of point-set `sample data`.

The semi-variogram measures the average dissimilarity between two variables, for example between the porosity at location $\mathbf{u}$ and at location $\mathbf{u} + \mathbf{h}$. Assuming stationarity, the semi-variogram $\gamma(Z(\mathbf{u}), Z(\mathbf{u} + \mathbf{h}))$ depends only on lag

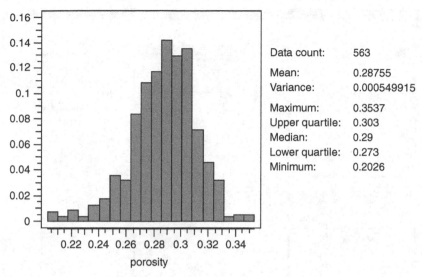

Data count:	563
Mean:	0.28755
Variance:	0.000549915
Maximum:	0.3537
Upper quartile:	0.303
Median:	0.29
Lower quartile:	0.273
Minimum:	0.2026

Figure 2.9 Histogram of the sample porosity

vector $\mathbf{h}$: $\gamma(Z(\mathbf{u}), Z(\mathbf{u} + \mathbf{h})) = \gamma(\mathbf{h})$. The experimental semi-variogram $\gamma(\mathbf{h})$ is computed as:

$$\gamma(\mathbf{h}) = \frac{1}{2N(\mathbf{h})} \sum_{\alpha=1}^{N(\mathbf{h})} [z(\mathbf{u}_\alpha) - z(\mathbf{u}_\alpha + \mathbf{h})]^2$$

where $z(\mathbf{u})$ is the (porosity) value at location $\mathbf{u}$ and $N(\mathbf{h})$ is the number of data pairs separated by vector $\mathbf{h}$.

In the rest of the text, the term *variogram* will liberally be used in place of the more precise *semi-variogram*. More background and references on variograms are provided in Section 3.5 and Chapter 5.

To compute the experimental variogram of porosity, open the variogram computation and modeling wizard by clicking *Data Analysis → Variogram*. The variogram computation wizard is thoroughly described in Chapter 5.

Select `sample data` and `porosity` as both *head* and *tail* properties. Selecting two different properties for head and tail would compute the cross-variogram between the two properties. On the next screen load the variogram parameters from file `variogram.par`, using the *Load Parameters...* button. Clicking *Next*, SGeMS computes the porosity variogram in two directions, azimuth 0° and azimuth 90° from North as well as an omni-directional variogram. All these parameters are discussed in Chapter 5.

The last screen of the wizard (see Fig. 2.10) shows the resulting variograms and provides controls to interactively fit a variogram model. The top left plot

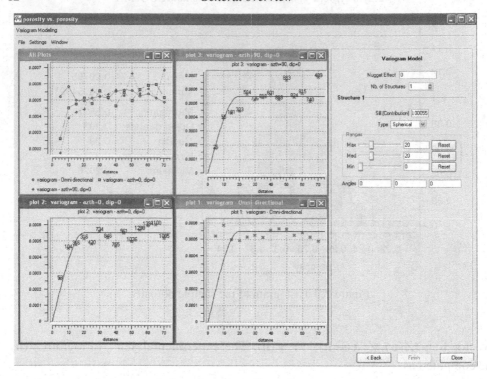

Figure 2.10 Last screen of the variogram modeling wizard

shows together all three experimental variograms. Each experimental variogram is also displayed in its own plot window. The controls in the right hand side panel interactively update the variogram model fit, overlaid on each plot.

An isotropic spherical variogram of range 20 and sill 0.00055 provides an acceptable fit to demonstrate the estimation/kriging process.

2.2.4 Creating a grid

The next step is to create the grid on which simple kriging will be performed. In this case a 3D Cartesian grid with $100 \times 130 \times 10$ cells is specified. Cartesian grids in SGeMS are 3D regular grids, i.e. all cells have orthogonal edges, and same dimensions. The grid is completely characterized by nine parameters (see Section 2.3 for more details):

- the number of cells in the X, Y and Z directions,
- the size of a cell in the X, Y and Z directions,
- the x, y, z coordinates of the origin of the grid.

Click *Objects → New Cartesian Grid* to open the grid creation dialog. Enter the dimensions of the grid, the coordinates of the grid origin, here (0,0,0), and the

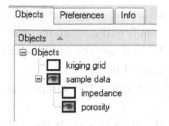

Figure 2.11 Object list after the Cartesian grid is created

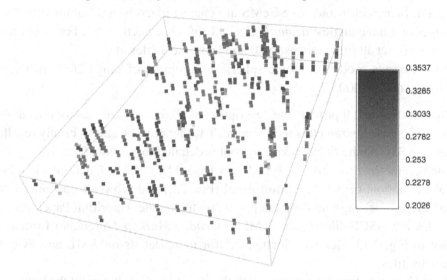

Figure 2.12 The 563 sample data and grid *kriging grid* in wire-frame

dimensions of each grid cell. Give a name for the new grid, `kriging grid` for example, and click *Create Grid* to create the grid. A new entry called `kriging grid` appears in the **Objects** panel of the Visualization Panel, as shown in Fig. 2.11.

The objects database now contains two objects: a point-set with the rock porosity property (and another property), and a Cartesian grid with no property attached yet.

Figure 2.12 gives the outline of the grid with the 563 sample data points inside.

2.2.5 Running a geostatistics algorithm

At this point, everything is ready to run the simple kriging algorithm: the sample data and the working grid are available to SGeMS and a variogram model has been created.

Select the kriging tool from the list in the Algorithm Panel. A form prompting for the kriging parameters appears below the algorithms list. The required parameters for simple kriging are:

- the name of the working grid, `kriging grid` in this case
- the name of the property which will contain the results of the kriging
- the simple kriging mean. We will use the sample mean $m^* = 0.2875$, which was computed during the EDA, see Fig. 2.9
- the name of the object containing the sample data: `sample data`, property `porosity`
- the size of the ellipsoid in which to search for conditioning data: data will be searched within a sphere of radius 80, more than twice the range of the variogram. Search ellipsoids in SGeMS are characterized by six parameters: three ranges and three angles: *azimuth*, *dip* and *rake*, see Section 2.5. For a sphere of radius 80, set all three ranges to 80 and leave the angles at $0°$
- the variogram model: isotropic spherical variogram of range 20, zero nugget effect and sill 0.00055.

Chapters 7, 8 and 9 provide the description of all the geostatistics tools available in SGeMS and the parameters they require. The theory of kriging is briefly recalled in Section 3.6 and the SGeMS kriging tool is detailed in Section 7.1.

The parameters can either be typed in or loaded from a file. Click the *Load* button at the bottom of the Algorithm Panel (Fig. 2.2) and browse to parameter file `kriging.par`, or drag-and-drop the parameter file on the Algorithm Panel. Parameter files are ASCII files in the XML (*eXtended Markup Language*) format, as shown in Fig. 2.13. Refer to Section 2.4 for more details on XML and SGeMS parameter files.

Once all parameters are entered, click the *Run Algorithm* button at the bottom of the Algorithm Panel. If some parameters are not correctly set, they are highlighted in red; a description of the error will appear if the mouse is left a few seconds on the offending parameter. Correct any error and click the *Run Algorithm* button again.

If kriging was run with the parameters shown on Fig. 2.13, the grid named `kriging grid` now contains two new properties: `estimated porosity`, and `estimated porosity_krig_var`, the associated kriging variance.

2.2.6 Displaying the results

The kriging results have been placed in property `estimated porosity`. Click on the plus sign before the `kriging grid` entry in the objects list to show the list of properties attached to that grid, and click in the square before `estimated porosity` to display the new property. Figure 2.14 shows a top view of grid `kriging grid` with property `estimated porosity` painted.

The Visualization Panel is an interactive 3D environment that enables objects to be viewed (e.g. grid `kriging grid`) from different angles, zoom-in/out, etc. Imagine a camera that can be moved freely in space around static objects. The Visualization Panel operates in two different modes: a camera mode, where the mouse

```
<parameters> <algorithm name="kriging" />

<Grid_Name value="kriging grid" />
<Property_Name value="estimated porosity" />

<Kriging_Type type="Simple Kriging (SK)" >
<parameters mean="0.27" />
</Kriging_Type>

<Hard_Data grid="sample data" property="porosity"/>
<Search_Ellipsoid value="80 80 80 0 0 0" />
<Min_Conditioning_Data value="0" />
<Max_Conditioning_Data value="20" />

<Variogram nugget="0" structures_count="1" >
<structure_1 contribution="0.003" type="Spherical" >
<ranges max="38" medium="38" min="38" />
<angles x="0" y="0" z="0" />
</structure_1>
</Variogram>

</parameters>
```

Figure 2.13 *Kriging* parameter file

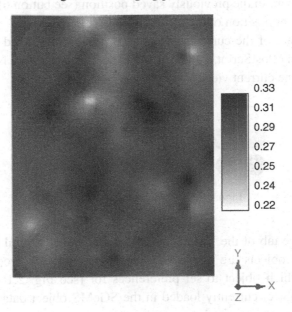

Figure 2.14 Kriging results – top view

controls the movements of the camera, and a selection mode, where the mouse can be used to select a displayed object. In camera mode, the mouse cursor looks like two curved arrows forming a circle, while in selection mode, the cursor is a standard pointer arrow. Press the *Escape* key to toggle between the two modes.

Controlling the camera

The camera is controlled with the mouse.

Rotation: left button click and drag in any direction to "turn" the objects in that direction (the camera moves in the opposite direction, giving the illusion that the objects were turned in the direction the mouse moved).

Translation: middle button click (or Shift + Left Click) and drag to translate the objects.

Zoom: mouse wheel (or Shift + Ctrl + Left Click) and drag to zoom in/out.

The bottom of the Visualization Panel provides several buttons (see Fig. 2.15) to further control the camera:

1. place the camera such that all displayed objects are visible
2. align the camera with the X axis
3. align the camera with the Y axis
4. align the camera with the Z axis (top view)
5. place the camera in the previously saved position (see button **6**)
6. save the current position of the camera
7. take a snapshot of the current view. The image can be saved in multiple formats, including PostScript, PNG or BMP. The image captured is exactly what is displayed in the current view.

Figure 2.15 Camera control buttons

The Preference tab of the Visualization Panel contains several controls to customize the way objects are displayed. The *Preferences for* drop-down list lets the user pick which object to set preferences for (see Fig. 2.16). There is one entry for each object currently loaded in the SGeMS object database, as well as a **<General>** preference panel.

The **<General>** preference panel has controls to:

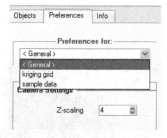

Figure 2.16 Selecting the object for which to set preferences

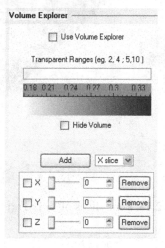

Figure 2.17 The Volume Explorer section of the preference panel for a Cartesian grid

- exaggerate the scale along the Z axis
- change the perspective mode: in *Perspective* view, the front face of a cube appears bigger than the back face, while in *Orthotropic* view distances and angles are conserved
- toggle the color of the background between white and black
- display the colorbar used for a given object. Although it is possible to display several objects at the same time, only one colorbar can be shown. By default the colorbar is placed on the right hand side of the view and can be moved with Alt + arrow keys or resized with Ctrl + arrow keys.

When visualizing the results of our simple kriging run, only the outer faces of the grid can be seen by default. However, the preference panel for `kriging grid` contains options to display only slices or part of the grid (using *volume rendering*). Select `kriging grid` in the *Preferences for* list. The preference panel is divided into three sections: General, Properties and Volume Explorer. The Volume Explorer section is shown in Fig. 2.17.

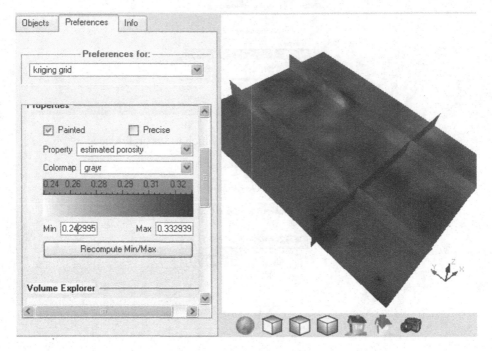

Figure 2.18 Four slices of `kriging grid`

We will first explore the grid by slicing it. Display `kriging grid` and check *Use Volume Explorer*. Check *Hide Volume* so that only slices are visible. Since we have not displayed any slice yet, nothing is visible in the view. By default only three slices can be viewed simultaneously, those orthogonal to X, Y and Z. Check the box next to a slice to display it and use the cursor to move the slice. If more slices are needed, select the orthogonal axis and click the *Add* button. Figure 2.18 shows four slices of `kriging grid`, two orthogonal to X, one orthogonal to Y and one orthogonal to Z.

Another useful application of the Volume Explorer is to hide parts of the grid. For example, it can be used to show only the grid cells with extreme porosity value and hide all the others. Uncheck *Hide Volume* and enter interval 0.255, 0.303 in the *Transparent Ranges* line. All cells with a porosity value between 0.255 and 0.303 are hidden as illustrated by Fig. 2.19.

It is possible to enter multiple intervals; each interval must be separated by a semi-colon. For example 0.2, 0.25; 0.3, 0.35 would hide all cells with a porosity value in the interval [0.2, 0.25] or the interval [0.3, 0.35].

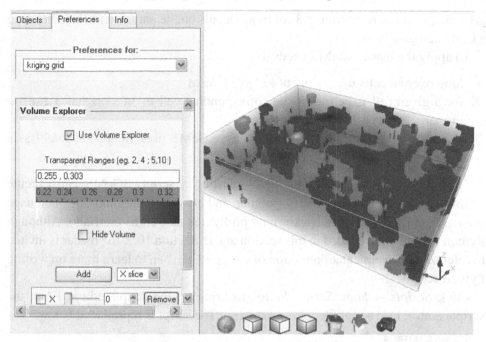

Figure 2.19 Volume rendering – all cells with a porosity value between 0.255 and 0.303 are hidden (almost completely transparent)

2.2.7 Post-processing the results with Python

Not all the cells of `kriging grid` may be of interest. The reservoir modeled by `kriging grid` features two main rock types: sand channels aligned in the North–South direction and background shales (see Fig. 2.5). All the 563 data points were sampled in the sand bodies, and were used to estimate porosity in the sand channels. This section demonstrates how to apply a mask and remove the non-sand channel cells from `kriging grid`.

File `mask.gslib` contains the mask data: it is a Cartesian grid in GSLIB format with a single property attached to each cell. The property is equal to 1 if the cell is in a channel, 0 otherwise. Click *Objects → Load Object* and browse to the mask file location (or drag-and-drop the file on the Visualization Panel). Since the GSLIB file format does not provide all the information required by SGeMS, a wizard prompting for additional information is popped up. The first screen of the wizard prompts for the type of the object, a Cartesian grid in this case. In the second screen, provide a name for the grid, e.g. `mask`, and indicate the number of cells in the grid ($100 \times 130 \times 10$), the size of each cell ($1 \times 1 \times 1$) and the origin of the grid (0,0,0). Refer to Section 2.3 for more details on the parameters characterizing a Cartesian grid in SGeMS.

To apply the mask, SGeMS needs to

1. loop over all cells $\mathbf{u}_1, \ldots, \mathbf{u}_n$ of `kriging grid`
2. for a given cell $\mathbf{u}_k$, check if the corresponding cell $\mathbf{u}_k$ in `mask` has a `facies` property equal to 0
3. if yes, remove the porosity value of $\mathbf{u}_k$. SGeMS uses value -9966699 as a no-data-value code.

This can easily be accomplished by writing a script. SGeMS can execute scripts written in Python, a very popular and powerful programming language (`www.python.org` provides background on Python as well as tutorials). Although Python is briefly introduced in this section and in Section 10.2, the reader is invited to refer to the Documentation section of `www.python.org` to learn more on writing Python scripts.

Click *Scripts* → *Show Scripts Editor* to bring up the scripts editor. From the scripts editor, load script file `apply_mask.py`. The script is reproduced below:

```
1  import sgems
2
3  mask=sgems.get_property('mask', 'facies')
4  porosity=sgems.get_property('kriging grid','estimated porosity')
5
6  for i in range(len(mask)):
7    if mask[i] == 0:
8      porosity[i]= -9966699
9  sgems.set_property('kriging grid','estimated porosity',porosity)
```

Following is the line-by-line explanation of the script.

Line 1 tells Python to load SGeMS specific commands (see Section 10.2), such as `sgems.get_property` and `sgems.set_property`.

Line 3 transfers the `facies` property of grid `mask` into an array called `mask`

Line 4 transfers the `estimated porosity` property of grid `kriging grid` into an array called `porosity`

Line 6 creates a loop, i will go from 0 to the size of array `mask` -1

Line 7 tests if the ith cell of grid `mask` is background shale

Line 8 if the ith cell is background shale, discard the estimated porosity value by setting it to SGeMS's not-a-number (NaN) code: -9966699

Line 9 transfers array `porosity` into property `estimated porosity` of grid `kriging grid`, overwriting the previous values in `estimated porosity`

Press the *Run* button at the bottom of the editor to execute the script. Any message from the script or error message is printed in the lower half of the editor, titled *Script Output Messages*.

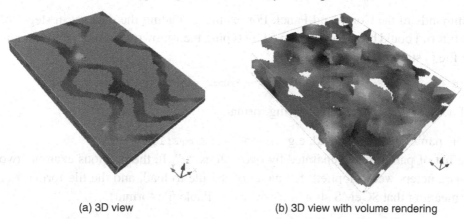

(a) 3D view   (b) 3D view with volume rendering

Figure 2.20 Final results of kriging, after applying the mask

Note that it is not required to use the SGeMS script editor. A script can be typed in more feature-rich editors and executed using command *Scripts → Run Script*. The results of the `apply_mask.py` script are shown in Fig. 2.20.

2.2.8 Saving the results

File → Save Project saves the project into a folder with the .prj extension, e.g. `tutorial.prj`. Folder `tutorial.prj` will contain one file for each object in the project, in SGeMS format (see Section 2.3 for a description on the SGeMS object file format). The project can later be loaded into SGeMS using *File → Open Project* or by dragging-and-dropping the project folder on the Visualization Panel.

Objects can also be saved individually: click *Object → Save Object* and provide a file name, the name of the object to save, e.g. `kriging grid`, and the file format to use (see Section 2.3).

2.2.9 Automating tasks

In the previous simple kriging run, SGeMS used at most 20 conditioning data to estimate the porosity at each cell (see parameter `Max_Conditioning_Data` of Kriging, Section 7.1). One may want to study the impact of that parameter on the results of kriging, for example varying the maximum number of conditioning data from 10 to 100 in increments of 5. Performing 19 simple kriging runs one at a time would be tedious.

SGeMS provides a solution to this problem through its command line interface. Most actions in SGeMS can either be performed with mouse clicks or by typing commands in the Command Panel. For example, loading the data set in step 1 of the tutorial could have been achieved by typing the following command (assuming the file is stored in D:/Tutorial):

```
LoadObjectFromFile D:/Tutorial/stanfordV_sample_data.gslib::All
```

Each command has the following format:

- the name of the command, e.g. LoadObjectFromFile
- a list of parameters, separated by two colons "::". In the previous example two parameters were supplied: the name of the file to load, and the file format All (meaning that SGeMS should try every available file format).

Every command performed in SGeMS, either typed or resulting from mouse clicks, is recorded to both the "Commands History" section of the Command Panel and to a file called sgems_history.log. Hence if one does not remember a command name, one can use the GUI to perform the corresponding action and check the command name and syntax in the command history.

It is possible to combine several commands into a single file and have SGeMS execute them all sequentially. For the sensitivity study example, one could write a *macro* file containing 19 kriging commands, each time changing the Max_Conditioning_Data parameter. Or better yet, write a Python script with a loop that executes the kriging command 19 times while updating the Max_Conditioning_Data parameter. File sensitivity_analysis.py, reproduced below, is an example of such a script.

```
1  import sgems
2
3  for i in range(0,19):
4    sgems.execute('RunGeostatAlgorithm
5    kriging::/GeostatParamUtils/XML::<parameters>
6    <algorithm name="kriging" />
7    <Grid_Name value="kriging grid"/>
8    <Property_Name value="titi"/>
9    <Kriging_Type type="Simple Kriging (SK)">
10   <parameters mean="0.27"/>
11   </Kriging_Type>
12   <Hard_Data grid="sample data" property="porosity"/>
13   <Search_Ellipsoid value="80 80 80 0 0 0"/>
14   <Min_Conditioning_Data value="0"/>
15   <Max_Conditioning_Data value="' + str(10+5*i) +'"/>
16   <Variogram nugget="0" structures_count="1">
17   <structure_1 contribution="0.003" type="Spherical">
18   <ranges max="38" medium="38" min="38"/>
19   <angles x="0" y="0" z="0"/>
20   </structure_1>
21   </Variogram>
22   </parameters>')
```

Although it looks lengthy, this script has only three statements: the `import` statement, the `for` loop, and the call to `sgems.execute` which executes the `RunGeostatAlgorithm`. The `RunGeostatAlgorithm` command is followed by all the parameters required by kriging on a single line, in XML format (see Section 2.4 for a description of SGeMS XML parameter files, and Section 7.1 for details on each parameter). The line of interest is line 15, where the value of parameter `Max_Conditioning_Data` is set to $str(10+5*i)$: Python evaluates $10 + 5 * i$ and turns the result into a string, which is concatenated with the rest of the parameters using the + operators.

2.3 Data file formats

In SGeMS, a Cartesian grid is defined by nine parameters:

- the number of cells nx, ny, nz in each direction
- the dimensions of a cell, $xsize, ysize, zsize$
- the coordinates of the origin.

This is illustrated in Fig. 2.21.

SGeMS supports two file formats by default to describe grids and sets of points: the GSLIB format and the SGeMS binary format.

The GSLIB file format

It is a simple ASCII format used by the GSLIB software (Deutsch and Journel, 1998, p.21). It is organized by lines:

- the first line gives a title. This line is ignored by SGeMS
- the second line is a single number n indicating the number of properties in the object, i.e. the number of columns of data

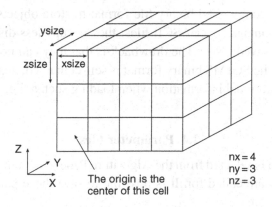

Figure 2.21 Parameters characterizing a Cartesian grid

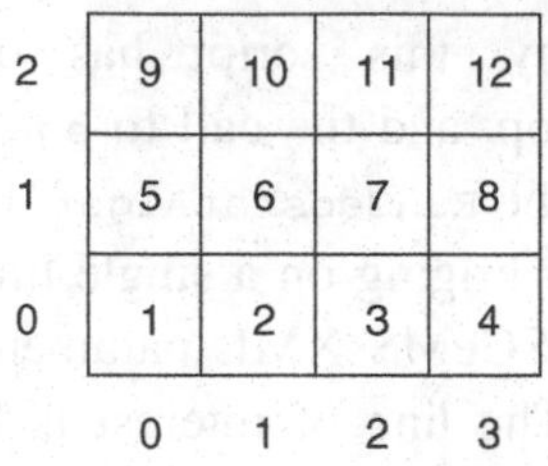

2	9	10	11	12
1	5	6	7	8
0	1	2	3	4

0   1   2   3

Figure 2.22 Implicit node coordinates of Cartesian grids, here a 4 × 3 grid. The numbers in the margin are the X and Y coordinates. The number in each cell indicates on which line the properties of that node should be entered

- the n following lines contain the names of each property (one property name per line)
- each remaining line contains the values of each property (n values per line) separated by spaces or tabulations. The order of the property values along each line is given by the order in which the property names were entered.

Note that the same format is used for describing both point-sets and Cartesian grids. When a GSLIB file is loaded into SGeMS the user has to supply all the information that is not provided in the file itself, such as the name of the object or the number of cells in each direction if it is a Cartesian grid.

In a Cartesian grid file, the X, Y, Z coordinates of each point are implicit. The first line of properties corresponds to the origin cell, which is the lower bottom left corner of the grid. Properties for node (i, j, k), the ith node in X, jth in Y and kth in Z (the origin is $i = 0, j = 0, k = 0$) are on line $k \times nx \times ny + j \times nx + i$, where nx, ny, nz are the number of cells in the X, Y and Z directions respectively. This is illustrated in Fig. 2.22.

The SGeMS binary file format

SGeMS uses an uncompressed binary file format to store objects. Binary formats have two main advantages over ASCII files: they occupy less disk space and they can be loaded and saved faster. The drawback is that they can not be easily edited with a text editor. The SGeMS binary format is self-contained, hence the user needs not provide any additional information when loading such a file.

2.4 Parameter files

When an algorithm is selected from the Algorithm Panel (see step 3 of Tutorial 2.2), several parameters are called for. It is possible to save those parameters to a file, and later retrieve them.

The format of a parameter file in SGeMS is based on the eXtended Markup Language (XML), a standard formatting language of the World Wide Web Consortium (www.w3.org). Figure 2.13 shows an example of such a parameter file.

In a parameter file, each parameter is represented by an XML element. An element consists of an opening and a closing tag, e.g. <tag> and </tag>, and one or several attributes. Following is the example of an element called algorithm which contains a single attribute "name":

```
<algorithm name="kriging"> </algorithm>
```

Elements can themselves contain other elements:

```
<Variogram nugget="0.1" structures_count="1">
    <structure_1 contribution="0.9" type="Spherical">
        <ranges max="30" medium="30" min="30"> </ranges>
        <angles x="0" y="0" z="0"> </angles>
    </structure_1>
</Variogram>
```

Here the element Variogram contains an element structure_1, which itself contains two elements: ranges and angles. Each of these elements have attributes: element Variogram has a nugget and a structures_count attribute, set here respectively to 0.1 and 1. Note that if an element only contains attributes, i.e. it has no nested elements, the closing tag can be abbreviated: in the previous example, element ranges only contains attributes and could have been written:

```
<ranges max="30" medium="30" min="30"/>
```

The /> sequence indicates the end of the element.

A SGeMS parameter file always has the two following elements.

- Element parameters. It is the root element: it contains all other elements.
- Element algorithm. It has a single attribute name which gives the name of the algorithm for which parameters are specified.

All other elements are algorithm-dependent and are described in Chapter 7 and Chapter 8.

Such an XML formatted parameter file has several advantages.

- Elements can be entered in any order.
- Comments can be inserted anywhere in the file. A comment block starts with <!-- and end with -->. They can span multiple lines, as shown by the following example.

```
<!-- An example of a comment block spanning
  multiple lines -->

<parameters> <algorithm name="kriging" />

  <!-- the name of the working grid -->
  <Grid_Name value="working_grid" />
</parameters>
```

2.5 Defining a 3D ellipsoid

Many algorithms in SGeMS require the user to specify a 3D ellipsoid, for example to represent a search volume through three anisotropy directions and affinity coefficients. In SGeMS, a 3D ellipsoid is represented by six parameters: the three radii of the ellipsoid r_{max}, r_{med}, r_{min}, and three angles, α, β, θ positioning the ellipsoid in space (see Figs. 2.23, 2.24 and 2.25).

Let (X, Y, Z) be the three orthogonal axes of a Cartesian coordinate system. The position of the ellipsoid is obtained by three successive rotations. Initially, before any rotation, the major axis Y' of the ellipsoid is aligned with Y, the medium axis X' with X, and the minor axis Z' with Z, see Fig. 2.23.

First rotation about Z' The ellipsoid is first rotated about axis Z', by angle $-\alpha$, where α is traditionally called *azimuth*, see Fig. 2.24. Looking in the direction of Z', the azimuth is measured counter-clockwise.

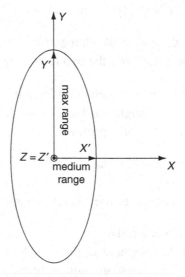

Figure 2.23 Starting point

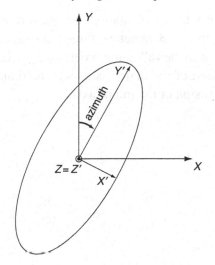

Figure 2.24 First rotation about Z'

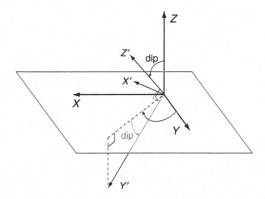

Figure 2.25 Second rotation about X'

Second rotation about X' The ellipsoid is then rotated about axis X', by angle $-\beta$, where β is called the *dip*, see Fig. 2.25. Looking in the direction of X', the dip is measured counter-clockwise.

Third rotation about Y' Last, the ellipsoid is rotated about axis Y' by angle θ, called the *plunge* or *rake* angle. Looking in the direction of Y', the plunge is measured clockwise.

The final transformation matrix, product of three rotations, is:

$$\mathbf{T} = \begin{bmatrix} \cos\theta & 0 & -\sin\theta \\ 0 & 1 & 0 \\ \sin\theta & 0 & \cos\theta \end{bmatrix} \begin{bmatrix} 1 & 0 & 0 \\ 0 & \cos\beta & \sin\beta \\ 0 & -\sin\beta & \cos\beta \end{bmatrix} \begin{bmatrix} \cos\alpha & \sin\alpha & 0 \\ -\sin\alpha & \cos\alpha & 0 \\ 0 & 0 & 1 \end{bmatrix}$$

Note that the order in which the rotations are performed is important. In the SGeMS parameter files, the six parameters, three radius and three angles, defining an ellipsoid must be given in the following order: r_{max}, r_{med}, r_{min}, azimuth, dip and rake; r_{max} is the dimension of the major axis, r_{med} the dimension of the medium axis, and r_{min} is the dimension of the minor axis.

3
Geostatistics: a recall of concepts

This user's manual is no place for another presentation of the theory of geostatistics. Many excellent books and reference papers are available to such purpose: Journel and Huijbregts (1978); Isaaks and Srivastava (1989); Cressie (1993); Wackernagel (1995); Goovaerts (1997); Deutsch and Journel (1998); Chilès and Delfiner (1999); Olea (1999); Lantuéjoul (2002); Mallet (2002). In this chapter we will only review the basic concepts and geostatistical principles underlying the algorithms offered in SGeMS. The more recent developments of multiple-point geostatistics are presented in relatively greater length because they are less known. Engineering-type presentations are preferred over more rigorous but less intuitive developments. These presentations point to programs coded into SGeMS whenever available.

A warning about references. We have limited citations to a few most relevant and easily accessible references. We give page numbers only to the three books by Goovaerts (1997), Deutsch and Journel (1998) and Chilès and Delfiner (1999). For an extensive list of references the reader may turn to the list proposed by Cressie (1993) and that of Chilès and Delfiner (1999).

Section 3.1 introduces the concept of random variable to model the uncertainty about a single variable. That concept is extended in Section 3.2 to a random function modeling the joint uncertainty about several interdependent variables distributed in space. The various possible outcomes of a random variable or random function are controlled by probability distribution functions made conditional to the data available; simulated outcomes can then be drawn from these conditional distributions, as discussed in Section 3.3. Conversely, a set of simulated outcomes can define a random function which is then seen as algorithm-driven, the algorithm being that used to generate these outcomes. At the core of any random function is a structural model that indicates how the various constitutive random variables relate to each other and to the data; inference of such model necessarily requires a prior decision of stationarity, as discussed in Section 3.4. In Section 3.5 we see that the structural model can be limited to a variogram-type relation between any two variables or involve many more than two variables at a time. In the latter case, inference of the corresponding multiple-point statistics calls for a training image. Section 3.6 presents the kriging paradigm which is at the origin of most geostatistical algorithms whether aimed at estimation or simulation. Section 3.7 introduces the theory underlying the multiple-point geostatistical algorithms. The traditional variogram-based simulation algorithms, *SGSIM, DSSIM* and *SISIM* are presented in Section 3.8. The two multiple-point simulation algorithms, *SNESIM* and *FILTERSIM*, are presented in Section 3.9. The nu/tau expression for combining probabilities conditional to different data events is given in Section 3.10; this non-traditional yet exact expression of the fully conditioned probability provides a useful separation of data information content and data redundancy.

3.1 Random variable

The conceptual model at the root of geostatistics, and for that matter of all of statistics and probability theory, is that of a random variable or random function. This is the model that allows making uncertainty assessment about an imperfectly known attribute or variable.

A deterministic variable takes only one outcome; that outcome is either known or unknown leaving no flexibility for uncertainty. Conversely, a random variable (RV) is a variable that can take a series of possible outcomes, each with a certain probability or frequency of occurrence (Goovaerts, 1997, p.63; Deutsch and Journel, 1998, p.11; Jensen *et al.*, 1997). A random variable is traditionally denoted with a capital letter, say, Z. Its possible outcomes are denoted with the corresponding small case letter, say, $\{z_i, i = 1, \ldots, n\}$ for a discrete variable with n outcomes, or $\{z \in [z_{\min}, z_{\max}]\}$ for a continuous variable valued in the interval bounded by a maximum and minimum value.

In the discrete case, to each outcome z_i is attached a probability value

$$p_i = \text{Prob}\{Z = z_i\} \in [0, 1], \quad \text{with:} \quad \sum_{i=1}^{n} p_i = 1. \tag{3.1}$$

In the continuous case, the distribution of probability values can take the form of

- a cumulative distribution function (cdf), pictured as a cumulative histogram, providing the probability for the RV not to exceed a given threshold value z, see Fig. 3.1a:

$$F(z) = \text{Prob}\{Z \leq z\} \in [0, 1] \tag{3.2}$$

- a probability density function (pdf), pictured as a histogram, defined as the derivative or the slope of the previous cdf at z-values where F is differentiable: $f(z) = dF(z)/dz$.

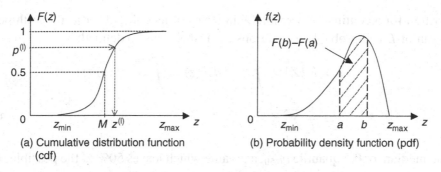

(a) Cumulative distribution function (cdf)

(b) Probability density function (pdf)

Figure 3.1 Probability distribution function

From such pdf or cdf, probability intervals can be derived, see Fig. 3.1b:

$$\text{Prob}\,\{Z \in (a, b]\} = F(b) - F(a) = \int_a^b f(z)dz. \qquad (3.3)$$

The key to a probabilistic interpretation of a variable z is the modeling of the distribution function, cdf or pdf, of the corresponding random variable Z. Note that such modeling does not mean necessarily fitting a parametric function to either the cdf or pdf; a series of classes with attached probability values is a valid model (Deutsch and Journel, 1998, p.16). That distribution function should account for all information available; it then provides all that is needed to quantify the uncertainty about the actual outcome of the variable z. For example,

- probability intervals can be derived as in Eq. (3.3);
- quantile values can be derived such as the 0.1 quantile or 1st decile:

$$q_{0.10} = F^{-1}(0.10) = z\text{-outcome value such that Prob}\,\{Z \le q_{0.10}\} = 0.10$$

- simulated values can be drawn by reading quantile values $z^{(l)}$ corresponding to a series of random numbers $p^{(l)}, l = 1, \ldots, L$ uniformly distributed in $[0, 1]$:

$$z^{(l)} = F^{-1}(p^{(l)}), \qquad l = 1, \ldots, L \qquad (3.4)$$

This process, called Monte Carlo drawing, ensures that the cdf of the L values $z^{(l)}$ will reproduce the Z-cdf $F(z)$, see Fig. 3.1a and Goovaerts (1997, p.351), Deutsch and Journel (1998, p.154). Conversely, a random variable Z can be modeled by the distribution of a number L of simulated values $z^{(l)}$ generated from a process initiated with equally probable uniform random numbers $p^{(l)}$: this is the concept of algorithm-driven random variable, see hereafter and Section 3.3, Deutsch (1994).

From the distribution of Z, specific moments or characteristic values can be derived such as,

- the mean m, or expected value of the RV Z, which can be retained as an estimate of the unknown value z, best in a least squared error sense. This mean is here written for a continuous variable Z in terms of its cdf, pdf, or as the arithmetic mean of L equiprobable realizations $z^{(l)}$ if the RV is so defined:

$$m = E\{Z\} = \int_{z_{\min}}^{z_{\max}} z\, dF(z) = \int_{z_{\min}}^{z_{\max}} z f(z) dz$$

$$= \frac{1}{L} \sum_{l=1}^{L} z^{(l)} \tag{3.5}$$

- the median, or 0.5 quantile $q_{0.50}$, a z-value which leaves 50% of the possible outcome values above it and 50% below it, see Fig. 3.1a. The median can be used as yet another estimate of the unknown value z, best in a least absolute error sense:

$$M = q(0.50) : \text{value such that Prob}\{Z \le M\} = 0.50 \tag{3.6}$$

- the variance which can be used as a single summary of the uncertainty around the mean estimate m:

$$\sigma^2 = \text{Var}\{Z\} = E\{(Z - m)^2\} = \int_{z_{\min}}^{z_{\max}} (z - m)^2 f(z) dz$$

$$= \frac{1}{L} \sum_{l=1}^{L} \left(z^{(l)} - m\right)^2. \tag{3.7}$$

Beware that the two most-used moments, mean and variance, generally do not suffice by themselves to define a distribution, hence to define probability intervals such as given in relation (3.3). Often a Gaussian-related distribution is adopted to provide the missing information. The problem is that errors associated to the various data integration processes involved in spatial interpolation are almost never Gaussian-distributed as opposed to direct errors due to measurement devices.

One definite advantage of an RV modeled through a set of L realizations $z^{(l)}$ is that probability intervals can be defined directly without going through any variance calculation. Also these probability intervals are independent of the particular estimated value retained, as opposed to the variance (Eq. (3.7)) which is specific to the mean estimate m. If one accepts that there is no unique "best in absolute" estimated value for any unknown, probability intervals and uncertainty measures should indeed be independent of the particular estimated value retained (Srivastava, 1987; Goovaerts, 1997, p.340; Isaaks and Srivastava, 1989).

Algorithm-driven random variable

One can argue that all of predictive geostatistics amounts to the determination of a probability distribution model, a model that accounts for *all* information available about the unknown value(s) z. A distribution model cannot be reduced to its mean and variance unless some two-parameter distribution is adopted; one must then question why mean and variance should be carefully determined if the far more consequential distribution type retained is not appropriate. Instead of determining mean and variance of the possible outcomes of an unknown, modern geostatistics aims at building a process (an algorithm) mimicking the data environment of that unknown; that algorithm then allows generating many (L) alternative outcomes of that unknown, although possibly not all of them. These L simulated realizations $z^{(l)}$, $l = 1, \ldots, L$, define an algorithm-driven random variable from which probability intervals for the unknown can be retrieved, as well as an estimated value which need not be the mean (Journel, 1994; Deutsch, 1994).

The number L of realizations can be as large as can be comfortably processed (Deutsch and Journel, 1998, p.133; Chilès and Delfiner, 1999, p.453). Note that a different set of L' realizations, with possibly $L = L'$, actually defines a different random variable. The number L and the specific L realizations retained are part of the defining algorithm.

This book and the SGeMS software provide tools for building these models of uncertainty. The details are in the "how to" and in the presentation, mostly graphical (maps), of the results.

Beware that a model of uncertainty is just that, a model, and there could be alternative models, each delivering possibly different results such as different estimates of the unknown, yet using the same original information (data) but in a different manner. There is no unique model of uncertainty, and most troublesome, there is neither a "best" model nor a fully objective model. We will return repeatedly to that point. Geostatistics, and for that matter all of probability theory, can only provide consistency with a prior model necessarily partly subjective, it cannot provide fully objective decisions (Goovaerts, 1997, p.442; Chilès and Delfiner, 1999, p.22; Matheron, 1978; Journel, 1994; Dubrule, 1994).

3.2 Random function

Most applications of geostatistics in the earth sciences involve mapping, which is the joint consideration of variables at several locations in space and/or time. Some of these variables are known through sampling; most others are unknown with varying degrees of uncertainty, they should therefore be modeled as random variables. However, we are not interested in evaluating each unknown independently of the others nearby; we are interested in an assessment of the *joint* spatial distribution of all unknowns, which is an assessment of their relation and connectivity in space. The uncertainty modeling should therefore consider all unknown variables together. The concept of a random function answers that requirement (Goovaerts, 1997, p.68; Chilès and Delfiner, 1999, p.12).

A random function (RF), denoted $Z(\mathbf{u})$, is a set of *dependent* random variables $\{Z(\mathbf{u}), \mathbf{u} \in S\}$, each marked with a coordinate vector $\mathbf{u}$ spanning a field or study area S. That field is typically a 3D physical volume, in which case $\mathbf{u} = (x, y, z)$ is the vector of the three Cartesian coordinates; the common notation (z) for the variable and the vertical coordinate does not usually pose a problem. The variable could also be time in which case $\mathbf{u} = t$, or it could involve both space and time as for atmospheric pressure in which case $\mathbf{u} = (x, y, z, t)$.

Just like a single random variable Z is characterized by a distribution function, say its cdf $F(z)$ for a continuous variable, an RF $Z(\mathbf{u})$ would be characterized by its multivariate distribution function:

$$\text{Prob}\{Z(\mathbf{u}) \leq z, \mathbf{u} \in S\} \tag{3.8}$$

a function of many parameters, any number N of locations $\mathbf{u}$ in S and the corresponding threshold values z possibly all different from one location to another.

The analytical expression of such multivariate distribution is impractical on a sizeable grid; there can be many millions of locations in a 3D grid. Exceptions are analytical distributions defined from a very small number of parameters, e.g. Gaussian-related (Anderson, 2003; Goovaerts, 1997, p.265; Chilès and Delfiner, 1999, p.404). But parameter-poor distributions are very specific in their properties, hence very restrictive. Last and not least, any advantage provided by an analytical definition diminishes when data locations and values are included into the field S, unless these data are consistent with the prior RF (unlikely in practice). This process of including into the RF $Z(\mathbf{u})$ random variables that are actually sampled is known as "data conditioning."

3.2.1 Simulated realizations

Just as a single RV can be defined by a finite set of simulated realizations, an RF $Z(\mathbf{u})$ is displayed and used through its realizations $\{z^{(l)}(\mathbf{u}), \mathbf{u} \in S\}$, $l = 1, \ldots, L$ (Lantuéjoul, 2002; Goovaerts, 1997, p.369; Chilès and Delfiner, 1999, p.449; Deutsch and Journel, 1998, p.119). In practice these realizations take the form of a finite number L of simulated maps, each providing an alternative, equally probable, representation of the unknown "true" map $z(\mathbf{u})$, $\mathbf{u} \in S$, see Fig. 3.2a. Any one specific such realization is denoted $z^{(l)}(\mathbf{u})$, $\mathbf{u} \in S$, where the upper script l indicates the realization number. A realization can be seen as a numerical model of the possible distribution in space of the z-values. That numerical model can be used to different purposes, including visualization and input to some transfer function representing the process under study, e.g. mining out the z-grade values.

Ideally, one would like to have access to an RF such that one of its realizations identifies the actual distribution of the true values $z(\mathbf{u})$, $\mathbf{u} \in S$; this is very unlikely if S contains millions or even only thousands of unknown values $z(\mathbf{u})$. In practice, this latter limitation is not a matter of concern if the L realizations available allow

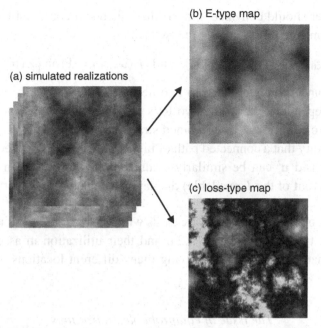

Figure 3.2 Simulated realizations and different estimation maps: (a) alternative equiprobable realizations of a random function; (b) E-type estimated map minimizing local squared error; (c) estimated map minimizing a specific loss function of the local error

a reasonable assessment of the consequent uncertainty on any processing of these unknown values. Recall the previous warning that an RF model is just a model, and asking that this model includes the unknown reality is naive.

It is the set of all such L simulated realizations, not any single realization, which provides an uncertainty assessment of the spatial distribution of the z-values over the study area. From a set of L realizations one could get the following.

- The probability distribution of the unknown $z(\mathbf{u})$ at any particular location $\mathbf{u}$ can be retrieved from the L simulated values $z^{(l)}(\mathbf{u})$ at that same location $\mathbf{u}$. This is much more than a mere estimation of $z(\mathbf{u})$, even if attached with an error variance since an error variance does not suffice to specify an error distribution. In a spatial interpolation setting, one could easily check through cross-validation that the cdf of the L simulated error values $\left[z^{(l)}(\mathbf{u}) - z^*(\mathbf{u})\right], l = 1, \ldots, L$, takes very different shapes depending on the data environment of location $\mathbf{u}$ and these shapes could be quite non-Gaussian.
- The probability that two nearby unknown values $z(\mathbf{u})$ and $z(\mathbf{u}')$ be simultaneously greater than any given threshold z_0 can be evaluated by the proportion of the L realizations which display simultaneously high simulated values at these two locations.

The reader should convince himself that this result could not be obtained, in general, from sole knowledge of the two cdfs

$$F(\mathbf{u}, z_0) = \text{Prob}\{Z(\mathbf{u}) \le z_0\} \quad \text{and} \quad F(\mathbf{u}', z_0) = \text{Prob}\{Z(\mathbf{u}') \le z_0\};$$

indeed the uncertainties related to two nearby values $z(\mathbf{u})$ and $z(\mathbf{u}')$ are not in general independent, hence the two RVs $Z(\mathbf{u})$ and $Z(\mathbf{u}')$ are not independent, and their two cdfs cannot be combined straightforwardly.

- The probability that a connected path of high z-values exists between two distant locations $\mathbf{u}$ and $\mathbf{u}'$ can be similarly evaluated by the proportion of simulated realizations (out of the L available) displaying such connected paths.

A great part of the SGeMS software deals with the generation of simulated realizations of the type shown in Fig. 3.2a, and their utilization in assessing spatial uncertainty, that is, uncertainty involving many different locations in space taken jointly.

The issue of equiprobable realizations

Through the cdf $F(z) = \text{Prob}\{Z \le z\}$ of a single analytically defined RV, one can define the probability of any outcome value z or class thereof. The case of an algorithm-driven RF is more delicate, if only because an analytical multivariate cdf characterizing that RF is rarely available; in practice it is never available after data conditioning. Indeed real and reasonably complex data are rarely perfectly consistent with a necessarily restrictive analytically defined RF model, hence conditioning that model to these data changes the model in an unpredictable way depending on the algorithm retained for the conditioning process.

In the vast majority of practical applications, the RF is defined through a finite number L of simulated realizations $\{z^{(l)}(\mathbf{u}), \mathbf{u} \in S\}, l = 1, \dots, L$. If any number $n \le L$ of realizations are *exactly* identical, these n realizations can be collated into a single one with probability n/L. However, in most practical applications no two realizations would be exactly identical, in which case all realizations are equiprobable, each with probability $1/L$.

How many realizations?

How many realizations (L) should be drawn from a given RF model (Chilès and Delfiner, 1999, p.453; Deutsch and Journel, 1998, p.133)? Since there is no reference model to approach, the number L should be chosen large enough to ensure stability of the results and small enough to allow the intended processing of the L simulated realizations.

Consider as "result" a specific function $\varphi\left(z^{(l)}(\mathbf{u}), \mathbf{u} \in S\right)$ built from any one simulated realization. The number L of realizations should be large enough such that statistics such as the variance of the L' values $\left\{\varphi\left(z^{(l)}(\mathbf{u}), \mathbf{u} \in S\right), l = 1, \ldots, L'\right\}$ stabilizes as L' increases towards L.

3.2.2 Estimated maps

There can be applications where values $z(\mathbf{u})$ are dealt with one at a time independently of the next one $z(\mathbf{u}')$, no matter how close the locations $\mathbf{u}$ and $\mathbf{u}'$ are. We will argue that those applications are few; even the most selective mining or environmental cleaning of single values $z(\mathbf{u})$ do account for close-by values $z(\mathbf{u}')$, e.g. for reason of cost and/or accessibility (Journel and Huijbregts, 1978; Isaaks and Srivastava, 1989). Notwithstanding, consider the derivation of a single-valued estimate $z^*(\mathbf{u})$ at each unsampled location $\mathbf{u}$.

Since the L simulated realizations of any single unsampled RV $z(\mathbf{u})$ are equiprobable, their point-wise arithmetic average provides a single estimated value, of least squared error-type, also called the E-type estimated value where E is short for "expected value", more precisely "conditional expectation" (Goovaerts, 1997, p.341; Deutsch and Journel, 1998, p.81):

$$z_E^*(\mathbf{u}) = \frac{1}{L} \sum_{l=1}^{L} z^l(\mathbf{u}). \tag{3.9}$$

The E-type map corresponding to the L realizations of Fig. 3.2a is shown in Fig. 3.2b. Note how much "smoother" the estimated map is compared to any one of the L simulated realizations. What is missing in an estimated map is the joint dependency between the many variables $Z(\mathbf{u}), \mathbf{u} \in S$, beyond their sharing common data. Far away from the data locations two estimated values would typically be identical; in the case of simple kriging they would be equal to the data mean value (Isaaks and Srivastava, 1989; Goovaerts, 1997, pp.130, 369). Clearly, geological or physical heterogeneity should not vanish just because there are no data nearby!

As a consequence, the probability that two nearby unknown values $z(\mathbf{u})$ and $z(\mathbf{u}')$ be simultaneously greater than any given threshold z_0 cannot be evaluated from an estimated map. More generally any assessment or estimation involving more than one location at a time should not be made based on an estimated map, particularly if that estimation is based on sparse data which is the case in many earth sciences applications.

In all rigor a set of estimated values should only be tabulated, not displayed as a map; a map entices the user to read relations between estimated z^*-values at different locations when these relations may not reflect that between the actual z-values. Only simulated values $z^{(l)}(\mathbf{u})$ should be mapped because a simulated realization is precisely a representation of the RF modeling the joint distribution in space of all random variables $Z(\mathbf{u}), \mathbf{u} \in S$. Beware, however, that if an estimated map is unique, there are many alternative equiprobable simulated realizations for any given data set. A simulation workflow forces the user to face uncertainty as represented by the L simulated realizations generated.

Alternative estimation criteria

Instead of the mean value defining the E-type estimated map as in expression (3.9), one could have retained the median $z_M^*(\mathbf{u})$ of the L simulated values $z^{(l)}(\mathbf{u})$ at each location $\mathbf{u}$. This would define an M-type estimated map, where each estimated value $z_M^*(\mathbf{u})$ has a 50% chance to be higher (or lower) than the actual unknown value. It can be shown that such M-type estimate is "best" in a least absolute error sense. In any particular application where single-location estimated values are needed, there is no a priori reason to minimize the squared error (e^2) or the absolute error $(|e|)$; one may want to minimize an application-specific loss function $\Phi(e)$, for example, in environmental applications using a non-linear function that penalizes underestimation of a lethal pollutant more than overestimation. Availability of the L simulated maps $\{z^{(l)}(\mathbf{u}), \mathbf{u} \in S\}$, $l = 1, \ldots, L$, allows considering such loss function-specific estimate, see Fig. 3.2c; Srivastava (1987), Goovaerts (1997, p.340).

Kriging could provide directly and faster an estimated map, similarbut not identical to the E-type map but it does not allow the flexibility to consider other types of estimates. In addition, the kriging variance map is an incomplete, when not misleading, measure of uncertainty as opposed to the distribution provided by the L simulated maps $\{z^{(l)}(\mathbf{u}), \mathbf{u} \in S\}$, $l = 1, \ldots, L$; see later discussion in Section 3.6.1, and Goovaerts (1997, p.180, Journel (1986), Chilès and Delfiner (1999, p.178).

3.3 Conditional distributions and simulations

As already stated, the main task of any probabilistic assessment is to build a model for the probability distribution of the unknown(s), either taken one at a time as displayed by the histogram of Fig. 3.1b, or altogether as displayed by the set of simulated realizations of Fig. 3.2a. The uncertainty of any unknown, or any number of unknowns taken jointly, necessarily depends on the amount and types of data available and their assumed relation to the unknown(s) considered. Take the simple example of a single unsampled continuous variable at location $\mathbf{u}$, and denote by $n(\mathbf{u})$ the set of data informing it. The relevant cdf providing an assessment of the uncertainty about the unsampled value $z(\mathbf{u})$ is specific to the location $\mathbf{u}$ and the data set $n(\mathbf{u})$ and is written (Chilès and Delfiner, 1999, p.380; Goovaerts, 1997, p.69):

$$F(\mathbf{u}; z|n(\mathbf{u})) = \text{Prob}\{Z(\mathbf{u}) \le z|n(\mathbf{u})\},$$

in words, the probability that the unknown $Z(\mathbf{u})$ be valued no greater than the threshold value z conditional to (knowing) the data set $n(\mathbf{u})$.

That conditional probability is, by definition, equal to the following ratio, with as numerator the probability of the event to be assessed $Z(\mathbf{u}) \leq z$ occurring jointly with the data event, and as denominator the probability of that data event occurring:

$$\text{Prob}\{Z(\mathbf{u}) \leq z|n(\mathbf{u})\} = \frac{\text{Prob}\{Z(\mathbf{u}) \leq z, n(\mathbf{u})\}}{\text{Prob}\{n(\mathbf{u})\}}. \qquad (3.10)$$

Expression (3.10) makes explicit the dependence of that cdf on the location $\mathbf{u}$, more precisely, the relation of that location with the $n(\mathbf{u})$ data retained. In all rigor, one should also make explicit the type, location and value of each datum constituting the set $n(\mathbf{u})$. Indeed if any aspect of that data set changes, the distribution (3.10) is changed.

Distribution (3.10) is called the conditional cumulative distribution function (ccdf) of the specific RV $Z(\mathbf{u})$ given the data set $n(\mathbf{u})$. When many unknowns $\{z(\mathbf{u}), \mathbf{u} \in S\}$ are jointly involved, the conditional probability required becomes multivariable; it is written as (Goovaerts, 1997, p.372; Anderson, 2003; Johnson, 1987):

$$\text{Prob}\{Z(\mathbf{u}) \leq z, \mathbf{u} \in S|n(\mathbf{u})\} = \frac{\text{Prob}\{Z(\mathbf{u}) \leq z, \mathbf{u} \in S, n(\mathbf{u})\}}{\text{Prob}\{n(\mathbf{u})\}}. \qquad (3.11)$$

Probability distributions of type (3.1) or (3.2) which are not made conditional to the data available are of little practical interest. Similarly, it is the moments of the conditional distributions which are of practical interest, and only those should be used for estimation. For example it is the mean of the ccdf $F(\mathbf{u}; z|(n))$ which should be used as the least squared error estimate of the unknown $z(\mathbf{u})$ at location $\mathbf{u}$, not the mean of the marginal distribution $F(z)$ as defined in Eq. (3.2) since that marginal distribution does not account for the specific dependence of location $\mathbf{u}$ with the data.

Similarly, the L realizations $\{z^{(l)}(\mathbf{u}), \mathbf{u} \in S\}$, $l = 1, \ldots, L$, displayed in Fig. 3.2a are useful only if they are outcomes of the multivariate probability distribution (Eq. (3.11)) conditioned to all relevant data available over the study field S (Goovaerts, 1997, p.372). For example, the arithmetic average of the L simulated values $z^{(l)}(\mathbf{u})$ at any given location $\mathbf{u}$ provides an estimate of the unknown $z(\mathbf{u})$ at that location; the cumulative histogram of these L simulated values provides a discrete representation of the ccdf (Eq. (3.10)), itself a measure of uncertainty about $z(\mathbf{u})$ (Journel and Huijbregts, 1978; Isaaks and Srivastava, 1989; Goovaerts, 1997, p.180).

Consider a specific zone or block V within S; each simulated map provides a simulated value for the average z-value over V denoted $z_V^{(l)}$. Then the histogram of the L simulated values $z_V^{(l)}, l = 1, \ldots, L$ provides a measure of uncertainty about the unknown average value z_V (Journel and Kyriakidis, 2004). Instead of a mere average taken over the zone V, any complex non-linear function of the z-values over all or part of the field S could be considered.

3.3.1 Sequential simulation

How does one go about building a complex distribution such as that in expression (3.11) involving jointly many unknowns and many data possibly of different types? This is indeed the main challenge of geostatistics. A solution to such a formidable task is provided by the "divide and conquer" paradigm.

1. Divide the problem into a series of easier problems involving only one unknown at a time, that is, address the easier problem of determining the ccdf (Eq. (3.10)) of each unknown $z(\mathbf{u})$. The task of recombining these elementary conditional probabilities accounting for spatial dependence under-lies the sequential simulation algorithm (Deutsch and Journel, 1998, p.125; Goovaerts, 1997, p.390; Chilès and Delfiner, 1999, p.462; Rosenblatt, 1952).
2. Divide the possibly large and complex data set $n(\mathbf{u})$ constituted of many dif-ferent data types into a set of smaller more homogeneous data sets $n_k(\mathbf{u})$, $k = 1, \ldots, K$, and address the easier problem of determining the K ccdfs $\text{Prob}\{Z(\mathbf{u}) < z | n_k(\mathbf{u})\}$ conditioned to each of the smaller data set $n_k(\mathbf{u})$. The nu/tau model presented in Section 3.9 addresses the problem of combining these K ccdfs into a single one of type (3.10) (Journel, 2002; Krishnan, 2004; Bordley, 1982; Polyakova and Journel, in press).

We will develop the first divide paradigm by considering the case of only three interdependent unknowns $z(\mathbf{u}_1)$, $z(\mathbf{u}_2)$, $z(\mathbf{u}_3)$, at different locations $\mathbf{u}_1$, $\mathbf{u}_2$, $\mathbf{u}_3$. These three interdependent variables could also relate to three different attributes, say the grades of three different metals. The generalization to more than three unknowns is immediate. The joint pdf of three random variables conditional to the same data set (n) can be decomposed as Goovaerts, 1997, p.376:

$$\text{Prob}\{Z(\mathbf{u}_1) = z_1, Z(\mathbf{u}_2) = z_2, Z(\mathbf{u}_3) = z_3 | (n)\} =$$
$$\text{Prob}\{Z(\mathbf{u}_1) = z_1 | (n)\} \cdot$$
$$\text{Prob}\{Z(\mathbf{u}_2) = z_2 | (n), Z(\mathbf{u}_1) = z_1\} \cdot \qquad (3.12)$$
$$\text{Prob}\{Z(\mathbf{u}_3) = z_3 | (n), Z(\mathbf{u}_1) = z_1, Z(\mathbf{u}_2) = z_2\}.$$

In words, the tri-variate joint pdf has been decomposed into the product of three univariate conditional pdfs, each involving only one variable, $Z(\mathbf{u}_1)$ first, then $Z(\mathbf{u}_2)$, last $Z(\mathbf{u}_3)$, but with increased data conditioning.

Provided that the problem of conditioning to the available data set (n) can be solved (see hereafter), each of these three single variable conditional pdfs can be determined. Decomposition (3.12) then allows the process of sequential simulation (Deutsch and Journel, 1998, p.125) more precisely, as follows.

- A value for $Z(\mathbf{u}_1)$ is drawn from the first pdf $\text{Prob}\{Z(\mathbf{u}_1) = z_1|(n)\}$, say that simulated value is $z_1^{(l)}$.
- Next a value for $Z(\mathbf{u}_2)$ is drawn from the second pdf
  $\text{Prob}\left\{Z(\mathbf{u}_2) = z_2|(n), Z(\mathbf{u}_1) = z_1^{(l)}\right\}$, say that value is $z_2^{(l)}$.
- Last a value for $Z(\mathbf{u}_3)$ is drawn from the third pdf
  $\text{Prob}\left\{Z(\mathbf{u}_3) = z_3|(n), Z(\mathbf{u}_1) = z_1^{(l)}, Z(\mathbf{u}_2) = z_2^{(l)}\right\}$, say that value is $z_3^{(l)}$.

  The three simulated values $z_1^{(l)}, z_2^{(l)}, z_3^{(l)}$, although drawn in sequence one after the other, stem from the joint tri-variate distribution conditional to the common data set (n).
- If another set of three simulated values is needed, one can repeat the process using different random numbers for the drawings.

The interdependence between the three variables $z_1^{(l)}, z_2^{(l)}, z_3^{(l)}$, has been taken into account by conditioning the simulation of each single variable to values of all previously simulated variables. We have traded the problem of simulating jointly many variables for that of simulating only one variable at a time but with an increasing conditioning data set, from (n) to $(n + 1)$ then $(n + 2)$. In practice, the problem created by the increasing data set size is solved by retaining into the conditioning data set of each variable only the closest or most related previously simulated values (Gómez-Hernández and Cassiraga, 1994; Goovaerts, 1997, pp.390, 400).

Retaining only the $n(\mathbf{u})$ closest data to inform any unknown location $\mathbf{u}$ amounts to an approximation of Eq. (3.12) since not all the previously simulated variables are taken into consideration. On the other hand, retaining only the closest data allows tighter local conditioning. A consequence of retaining only the closest previously simulated values is that the sequence along which the nodes are visited matters. That sequence is called the simulation path, it is usually random to avoid artifacts (Daly and Verly, 1994).

The joint-pdf in Eq. (3.12) using the sequence $\{\mathbf{u}_1, \mathbf{u}_2, \mathbf{u}_3\}$ and retaining only one previously simulated value in addition to the original (n) becomes, with $\mathbf{u}_2$ being closer to $\mathbf{u}_3$ than $\mathbf{u}_1$:

$$\begin{aligned}
\text{Prob}\{Z(\mathbf{u}_1) = z_1, Z(\mathbf{u}_2) &= z_2, Z(\mathbf{u}_3) = z_3|(n)\} \approx \\
&\text{Prob}\{Z(\mathbf{u}_1) = z_1|(n)\} \cdot \\
&\text{Prob}\{Z(\mathbf{u}_2) = z_2|(n), Z(\mathbf{u}_1) = z_1\} \cdot \\
&\text{Prob}\{Z(\mathbf{u}_3) = z_3|(n), Z(\mathbf{u}_2) = z_2\}.
\end{aligned} \tag{3.13}$$

Another realization could be obtained by changing the uniform random numbers used for the drawing and/or changing the sequence in which the locations $\{\mathbf{u}_1, \mathbf{u}_2, \mathbf{u}_3\}$ are visited and simulated.

3.3.2 Estimating the local conditional distributions

The critical step in sequential simulation consists of estimating at each location $\mathbf{u}$ along the simulation path the conditional distribution given a specific conditioning data set $(n(\mathbf{u}))$. There have been essentially two avenues for approaching the determination of conditional pdf Prob $\{Z(\mathbf{u}) = z | n(\mathbf{u})\}$ the single variable $Z(\mathbf{u})$, both calling for a multiple-point (mp) RF model.

1. The traditional 2-point statistics approach consists of evaluating the relation of the single unknown $Z(\mathbf{u})$ with one datum $Z(\mathbf{u}_\alpha)$ at a time; thus no more than 2 locations or 2 variables are involved at any time. Such relation typically takes the form of a covariance/correlation or, equivalently, a variogram; these are 2-point statistics. Then calling on a prior multiple-point model requiring only 2-point stats for its calibration, the previous conditional pdf Prob $\{Z(\mathbf{u}) = z | (n)\}$ is determined through some form of kriging. Examples of such simple RF models that can be calibrated from only 2-point statistics are:
   - the multivariate Gaussian model underlying the sequential Gaussian simulation algorithm (SGeMS program *SGSIM*, Section 8.1.2), Goovaerts (1997, p.380); Anderson (2003); Chilès and Delfiner (1999, p.462); Gómez-Hernández and Journel (1993)
   - the truncation at order two of an expansion of the exact conditional probability in the discrete case. Such truncation underlies the indicator simulation algorithm (SGeMS program *SISIM*), see decomposition (3.12) and Section 3.8.4, Goovaerts (1997, p.393), Journel and Alabert (1989).
   In short, a 2-point statistics approach aims at dividing the data set (n) into single locations or variables. First each single data variable is related to the single unknown $(1 + 1 = 2$-point statistics), then these elementary results are pieced together through kriging using some simple prior multiple-point (mp) probabilistic model. The results are no better than this prior mp model, be it Gaussian-related or algorithm-driven.
2. The second approach avoids such extreme division of the conditioning data event. Instead it consists of using an explicit multiple-point (mp) model, which allows considering the (n) data altogether, or a set $n(\mathbf{u})$ of neighbor data. The necessary $n(\mathbf{u}) + 1$ multiple-point statistics are lifted from replicates of the $n(\mathbf{u})$-data event found in a visually explicit training image (Ti) (Guardiano and Srivastava, 1993; Srivastava, 1994; Strebble, 2000, 2002; Zhang *et al.*, 2006). The results of such explicit mp geostatistics application are no better than the

prior model implicit to the Ti used. The consideration of training images, if available, allows making use of mp structural information much beyond the variograms of these Tis.

A training image is a representation of how z values are jointly distributed in space (Farmer, 1992; Strebelle, 2002; Journel, 2002; Zhang, 2006). A training image (Ti) is essentially an unconditional realization of an RF model $Z(\mathbf{u})$, that is a prior conceptual depiction of the distribution in space of z-values, a depiction that need not honor at their location any of the data values included in the set (n). The joint distribution in space of the actual unknown values $\{z(\mathbf{u}), \mathbf{u} \in S\}$ is assumed to "look like" the Ti but would also honor the data (n). The role of an mp simulation is strictly one of data conditioning, "morphing" the Ti to honor the conditioning data (n). A 2-point simulation aims at generating simulated realizations that honor the data and a variogram model. An mp simulation aims at generating simulated realizations that honor the data and the multiple-point structures present in the training image.

The necessity of an mp model

It is important to understand that there cannot be any probabilistic estimation or simulation without the necessary multiple-point (mp) statistics linking the data taken altogether to the unknown(s) (Journel, 1994). Those mp statistics are either delivered explicitly through an analytical multivariate model or a training image, or they are implicitly provided by the specific simulation algorithm retained. Traditional algorithms that call for input of only 2-point statistics (variograms) adopt implicitly the higher order statistics in-built into the simulation algorithm retained, and these are most often of high entropy character. High or maximum entropy leads to maximizing disorder beyond the input variogram model(s) (Goovaerts, 1997, p.335; Journel and Alabert, 1989; Journel and Deutsch, 1993). An implicit model that maximizes disorder beyond specified 2-point statistics is no less a model than a training image with its specific (lower entropy) structures and patterns much beyond the reach of a mere variogram. Also it can be argued that a high entropy model is often inappropriate for an earth sciences application where complex curvilinear structures simultaneously involving many more than 2 locations in space are known to exist, even if these structures are not immediately apparent from the limited local data available. The 2-point statistics-based realizations are consistent with the implicit maximum entropy hypothesis beyond the input variogram model. If, however, one has any inkling of the existence of definite structures or patterns, this is precious structural information that must be accounted for in addition to the local data in the exercise of building alternative representations of the true image (Journel and Zhang, 2006). Two-point statistics, covariance or variogram, do not suffice to carry such mp information.

3.4 Inference and stationarity

The concept of stationarity is at the basis of all probabilistic inference: you try to associate the (data) environment of any unknown value to "similar" environments for which you know the outcome of the variable, this allows you to make prediction of the unknown from the known outcomes. The critical decision is that of similarity of the data environments, a decision which is never fully objective even though it defines the probabilistic model and thus impacts critically the predictions made (Goovaerts, 1997, p.70; Chilès and Delfiner, 1999, p.16; Deutsch and Journel, 1998, p.12; Wackernagel, 1995; Journel and Huijbregts, 1978).

Consider the most elementary problem of inferring the possible outcomes of a single unknown value z, which is inference of the distribution of the corresponding random variable Z. To be more specific, consider a petroleum reservoir and say that z is the unsampled porosity value at a given location $\mathbf{u}$, in which case the corresponding RV is denoted $Z(\mathbf{u})$. Many alternative decisions, all somewhat subjective, are possible.

- One may associate broadly the environment of the unknown $z(\mathbf{u})$ to the entire reservoir S which includes the location $\mathbf{u}$; in which case the distribution of $Z(\mathbf{u})$ could be inferred from the histogram of all samples $z(\mathbf{u}_\alpha)$ available in S whether the sample location $\mathbf{u}_\alpha$ belongs or not to the lithofacies prevailing at location $\mathbf{u}$. The decision of stationarity then encompasses the whole reservoir.
- If one knows that the unsampled location $\mathbf{u}$ is within a sand facies, it would make sense to restrict the previous histogram to only those samples known to have been collected in a sand facies. The decision of stationarity is now restricted to the sand facies. There is one caveat, however: there should be enough sand samples in S for their porosity histogram to be deemed representative; if not, one may have to pool samples from different lithofacies into the same histogram, or consider using sand porosity samples coming from reservoirs other than S but deemed similar. Thus, the decision of stationarity is necessarily subjective, conditioned in particular by data availability; that decision will change as the reservoir S matures, becoming better known and sampled. Yet, a different decision of stationarity implies a different probabilistic model, different data and different estimation results. The reservoir S is the same, it is our model of it which has changed.
- Consider the favorable case of a well sampled reservoir S where enough sand porosity samples are available to build a histogram deemed reliable. Should that histogram be considered as the probabilistic model for the location-specific RV $Z(\mathbf{u})$? Among those sand samples, one may want to give more weight to samples $z(\mathbf{u}_\alpha)$ at locations $\mathbf{u}_\alpha$ closer to the unsampled location $\mathbf{u}$, and also give more

weight to isolated sample locations as opposed to clustered sample locations to reduce the impact of preferential over-sampling (data clustering) in certain zones. This suggestion is at the basis of the concept of kriging (Krige, 1951; Matheron, 1970; Journel and Huijbregts, 1978). In the process of kriging, the Euclidean distance between any two (sand) locations $|\mathbf{u} - \mathbf{u}'|$ is replaced by a variogram distance $\gamma(\mathbf{u}, \mathbf{u}')$ read from a variogram model, itself inferred from the sand porosity samples. Inference of that variogram model calls for extending the decision of stationarity to pairs of sample values $z(\mathbf{u}_\alpha)$ and $z(\mathbf{u}_\beta)$ separated by approximately the same distance vector $\mathbf{h} = \mathbf{u}_\beta - \mathbf{u}_\alpha$ (Deutsch and Journel, 1998, p.43; Goovaerts, 1997, p.50). A model of isotropy, whereby only the modulus of vector $\mathbf{h}$ is retained, actually corresponds to yet another extension of the stationarity decision allowing the pooling of sample pairs with the same separation distance $|\mathbf{u}_\beta - \mathbf{u}_\alpha|$ irrespective of the direction of the vector $\mathbf{u}_\beta - \mathbf{u}_\alpha$. In the end, not just the kriging is important, but also the series of decisions of stationarity which allowed implementing that kriging, in our case stationarity of porosity in sand, then stationarity (of order 2) allowing inference of the variogram model, last and not least the local stationarity decision used to specify how far away from location $\mathbf{u}$ should one go (within sand) to define the data event $n(\mathbf{u})$ informing that location $\mathbf{u}$.

In many applications involving subsurface deposits where local data are sparse, inference of a variogram is difficult particularly if the decision of stationarity restricts samples to one specific facies, rock type or sub zone. Yet without a variogram there is no kriging, hence no traditional geostatistics. In the presence of sparse data, the necessary variogram is often borrowed from deposits or outcrops that are analogous to the spatial phenomenon under study, or it is simply drawn to reflect the geologist's appreciation of correlation ranges. The problem is that the spatial variability of many earth systems do not lend themselves well to a variogram characterization: very different patterns of variability may share the same variogram, see Fig. 3.3, Fig. 3.4, and Strebelle (2000) and Caers (2005). As for borrowing or drawing a variogram to depict the spatial variability of a variable $z(\mathbf{u})$, $\mathbf{u} \in S$, why not borrow or draw a more relevant conceptual image of that variability? Geologists do not think in terms of variograms or covariance matrix, their expertise is best expressed in the form of pictures, sketches, cartoons, as is evident to anyone opening a structural geology book. One could consider an outcrop photograph or a geological sketch as a training image, which is a realization of a random function model. The corresponding decision of stationarity is that the actual data environment of the unsampled value $z(\mathbf{u})$ in S, has approximate replicates over that training image (Strebelle, 2002; Journel and Zhang, 2006).

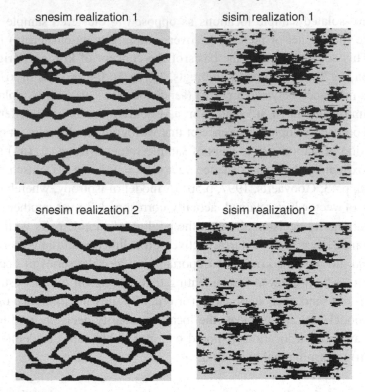

Figure 3.3 The two *SNESIM* generated images have more continuous and smoother features than the *SISIM* generated images, nevertheless they share the same variogram model

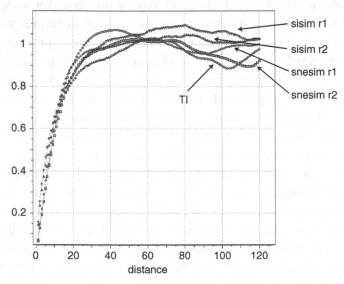

Figure 3.4 Reproduction of the E–W training images variogram by the two *SNESIM* and two *SISIM* generated images shown in Fig. 3.3

The subjectivity of model decision

It does take some effort for accepting the fact that stationarity, which is theoretically defined as invariance by translation of spatial statistics (Goovaerts, 1997, p.70; Chilès and Delfiner, 1999, p.16), is a modeling decision, not some hypothesis or property of the data which could be tested (Matheron, 1978; Journel, 1986). Such a decision is necessarily subjective and can only be judged after the fact by evaluating whether or not the resulting model has helped in achieving the task at hand. Stationarity is at the source of any prediction process, it defines the repetition process that provides replicates; without such repetition there is no inference possible. Unless a prior model is given fully determined, stationarity is a necessary decision that allows building the random function model and the inference of its characteristic moments. Accepting a fully defined model, such as the independence model, or a Gaussian model with a given covariance, or accepting a specific training image, amounts to different decisions of stationarity.

- In the case of a training image, its selection represents a decision of stationarity, which allows its scanning to find replicates of any specific data event and retrieval of the corresponding replicates of the central value informed by that data event. The histogram of these central value replicates is then taken as the conditional distribution of the actual unknown. This is the inference process underlying the *SNESIM* mp simulation algorithm; see Section 3.7, and Guardiano and Srivastava (1993) and Strebelle (2000).
- Accepting a Gaussian model allows building a kriging system to retrieve by kriging the two conditional moments (mean and variance) which suffice to specify the Gaussian conditional distribution; see Section 3.6, and Goovaerts (1997, p.266), Anderson (2003) and Chilès and Delfiner (1999, p.381).

A priori, one decision is no better than another. There is no universality attached to the Gaussian model or any theoretical aura attached to the task of solving a system of equations (kriging) as opposed to the more trivial task of scanning a training image (Ti) for replicates. Conversely, the patterns described by the Ti retained may not be the ones relevant to the actual deposit S. Adopting a wrong Ti may lead to severe errors, all the more dangerous that one is comforted by the final numerical representation of S fitting both the data and one's prior (and possibly erroneous) vision of the structures of variability. Hence, we insist on the absolute necessity of considering alternative different structural models, whether these are explicit training images or implicit models anchored on variogram models; these different structural models should reflect the range of possible different (geological) scenarios for the spatial variability.

Which model, 2-point or mp?

The better model is that which delivers the "deemed" better result: did mimicking the training image patterns yield a "more" satisfactory result than the simplified structures result yielded by variogram-based geostatistics? Again, the final judgment is necessarily case- and application-dependent and is in part subjective.

One could leave aside some critical data, of a global nature as opposed to the local data used for conditioning, and check which of the final numerical representations of S fits best these "check" data. Examples of such test data could be some of the production data in a producing hydrocarbon reservoir or mining deposit, they could be a subset of "ground truth" data in an environmental application (Caers and Hoffman, 2006; Goovaerts, 1997, p.105; Journel and Zhang, 2006).

There is place in a geostatistical toolbox for both sets of algorithms, 2-point and mp statistics are complementary: SGeMS proves it.

3.5 The variogram, a 2-point statistics

The main tool for traditional geostatistics, and for that matter, most statistical prediction algorithms, is the variogram or its equivalent (when defined) the covariance. Consider a stationary random function $Z(\mathbf{u})$, and any two of its random variables $Z(\mathbf{u})$ and $Z(\mathbf{u} + \mathbf{h})$ separated by vector $\mathbf{h}$. The relation between these two RVs is characterized by any one of the following 2-point statistics, functions of the separation vector $\mathbf{h}$ (Anderson, 2003; Goovaerts, 1997, p.28; Journel and Huijbregts, 1978; Matheron, 1970):

- the covariance:

$$C(\mathbf{h}) = E\big\{ [Z(\mathbf{u}) - m][Z(\mathbf{u} + \mathbf{h}) - m] \big\} \qquad (3.14)$$

- the correlogram, or coefficient of correlation:

$$\rho(\mathbf{h}) = C(\mathbf{h})/C(0) \in [-1, +1]$$

- the variogram:

$$2\gamma(\mathbf{h}) = E\left\{ [Z(\mathbf{u} + \mathbf{h}) - Z(\mathbf{u})]^2 \right\}$$
$$= 2[C(\mathbf{h}) - C(0)], \text{ if } C(\mathbf{h}) \text{ exists,}$$

where $m = E\{Z(\mathbf{u})\}$, $C(0) = \sigma^2 = \text{Var}\{Z(\mathbf{u})\}$ are the stationary marginal 1-point statistics.

Any of these 2-point moments can be inferred by the corresponding experimental statistics, say from $n(\mathbf{h})$ pairs of data $z(\mathbf{u}_\alpha + \mathbf{h})$, $z(\mathbf{u}_\alpha)$, $\alpha = 1, \ldots, n(\mathbf{h})$ approximately distant of $\mathbf{h}$. For example, the experimental variogram is given by (Goovaerts, 1997, p.28; Wackernagel, 1995; Isaaks and Srivastava, 1989):

$$2\gamma^*(\mathbf{h}) = \frac{1}{n(\mathbf{h})} \sum_{\alpha=1}^{n(\mathbf{h})} [z(\mathbf{u}_\alpha + \mathbf{h}) - z(\mathbf{u}_\alpha)]^2. \tag{3.15}$$

Modeling

In practice, available information only provides enough data pairs for a few distances $|\mathbf{h}|$ and along a few directions. However, that statistics, say $\gamma(\mathbf{h})$, is needed for all vectors $\mathbf{h} = \mathbf{u} - \mathbf{u}_\alpha$ linking an unsampled location $\mathbf{u}$ to any nearby datum location $\mathbf{u}_\alpha$; thus there is a need to interpolate/extrapolate experimental statistics such as $\gamma^*(\mathbf{h})$ into a model $\gamma(\mathbf{h})$ available for all $\mathbf{h}$. Because covariance and variogram are used to calculate variances and these variances are non-negative, not all analytical function $g(\mathbf{h})$ can be used as a covariance or a variogram model (Chilès and Delfiner, 1999, p.59; Goovaerts, 1997, p.87; Journel and Huijbregts, 1978; Christakos, 1984). Positive linear combinations of basic acceptable (licit) models $g(\mathbf{h})$ are also acceptable, this allows defining a large family of acceptable covariance/variogram models sufficient for most practical studies. SGeMS allows consideration of any positive linear combination of the three most commonly used basic variogram models, the spherical, exponential and Gaussian models introduced in Chapter 5 (Deutsch and Journel, 1998, p.25).

The reader is referred to Goovaerts (1997, p.87) and other relevant papers and publications (Matheron, 1962, 1962; David, 1977; Journel and Huijbregts, 1978; Journel and Froidevaux, 1982; Chauvet, 1982; Cressie, 1993; Yao and Journel, 1998) for the practice of modeling experimental variograms, a sometimes delicate task in presence of directional anisotropy, sparse data and prior non-quantitative information.

Cross-variogram

In the previous expressions (3.14) and (3.15), the two RVs may relate to two different attributes, say $Z_1(\mathbf{u})$ is porosity at location $\mathbf{u}$ and $Z_2(\mathbf{u} + \mathbf{h})$ is seismic impedance measured at location $\mathbf{u} + \mathbf{h}$. The corresponding 2-point statistics is then a cross-variogram defined as (Goovaerts, 1997, p.46, Chilès and Delfiner, 1999, p.328, Journel and Huijbregts, 1978; Wackernagel, 1995):

$$2\gamma_{12}(\mathbf{h}) = E\{[Z_1(\mathbf{u} + \mathbf{h}) - Z_1(\mathbf{u})][Z_2(\mathbf{u} + \mathbf{h})) - Z_2(\mathbf{u})]\}. \tag{3.16}$$

In presence of only two different attributes Z_1, Z_2, one must model a matrix of four (cross) covariance functions, $C_{11}(\mathbf{h})$, $C_{12}(\mathbf{h})$, $C_{21}(\mathbf{h})$, $C_{22}(\mathbf{h})$, or only three (cross) variogram functions $\gamma_{11}(\mathbf{h})$, $\gamma_{12}(\mathbf{h}) = \gamma_{21}(\mathbf{h})$, $\gamma_{22}(\mathbf{h})$, under restrictive conditions of positive definiteness, see Goovaerts (1997, p.108). In presence of N different attributes $Z_1, \ldots, Z_N$, there would be N^2 (cross) covariance functions or $N(N + 1)/2$ (cross) variogram functions to model! Subsurface data are rarely enough to allow inference of statistics involving more than $N = 2$ attributes simultaneously, even if these statistics are only 2-point statistics involving only 2 space locations at a time, $\mathbf{u}$ and $\mathbf{u} + \mathbf{h}$.

Multiple-point statistics

To characterize the relation between two patterns of data, say, n_1 data on attribute Z_1: $\{z_1(\mathbf{u} + \mathbf{h}_\alpha); \alpha = 1, \ldots, n_1\}$ and n_2 data on attribute Z_2: $\{z_2(\mathbf{u}' + \mathbf{h}'_\beta);$ $\beta = 1, \ldots, n_2\}$, one would need much more than cross-covariances or cross-variograms. One needs in all rigor the **joint** distribution of the $(n_1 + n_2)$ RVs $Z_1(\mathbf{u} + \mathbf{h}_\alpha), Z_2(\mathbf{u}' + \mathbf{h}'_\beta); \alpha = 1, \ldots, n_1; \beta = 1, \ldots, n_2$ (Goovaerts, 1997, p.72). No experimental data would be ever enough to infer such multiple-variable, multiple-point statistics; not to mention the nightmare of their modeling. There are two escape avenues.

1. Assume a parameter-poor random function model $\{Z_1(\mathbf{u}), Z_2(\mathbf{u}')\}$ fully defined from a few low-order statistics that can be inferred from data. Most often such models are related to the multivariate Gaussian model fully characterized by the sole covariance matrix $[C_{ij}; i, j = 1, \ldots, N]$, with $N = 2$ for the example above (Anderson, 2003; Goovaerts, 1997, p.265).
2. Build training images depicting the relation in space of the two variables $z_1(\mathbf{u})$ and $z_2(\mathbf{u}')$. These training images should reflect whatever physics or geology is known to control the joint spatial distributions of these two variables (Strebelle, 2000; Arpat, 2004; Zhang, 2006; Journel and Zhang, 2006).

In both cases, most of the structural (n-point statistics) information capitalized upon for estimation or simulation of an unsampled value (or set of values) is coming not from the data but from the model, multivariate Gaussian in case 1, the training image in case 2.

It would be a severe error, though both naive and common, to believe that one could do away with models falling in either of the two previous categories. The reason is that, whenever maps of estimated or simulated values are used, one necessarily draws from such maps much more than the 2-point or lower order statistics actually modeled from data; this "much more" comes from the n-point statistics of the RF model, whether Gaussian-related or training image-based (Journel, 1994).

3.6 The kriging paradigm

Kriging has been historically at the source of acceptance of geostatistics (Krige, 1951; Matheron, 1970; Journel and Huijbregts, 1978); it remains a major data integration tool and is used in most geostatistical estimation and simulation algorithms. In its simplest indicator kriging form with a single (normal) equation, it identifies Bayes relation and the very definition of a conditional probability (Journel, 1983).

Kriging is in essence a generalized linear regression algorithm (Goldberger, 1962; Luenberger, 1969), extending the data-to-unknown correlation to data-to-data correlation through a non-diagonal kriging matrix. It is a regression with non-independent data: actually it can be shown that kriging consists, first of de-correlating the data by defining linear combinations of the original data that are orthogonal for a given covariance/variogram model, then of a traditional linear regression from these "independent" data transforms (Journel, 1989).

The practice of geostatistics in very diverse earth sciences fields has led to a large number of variants from the theoretically rigorous yet simplest "simple kriging."

3.6.1 Simple kriging

Because kriging is at the source of so many geostatistical algorithms it is worth briefly recalling here the basic simple kriging (SK) system, then its multiple variants (Goovaerts, 1997, p.127; Deutsch and Journel, 1998, p.77; Chilès and Delfiner, 1999, p.154).

Consider within a stationary field S the estimation of an unsampled value $z(\mathbf{u})$ from $n(\mathbf{u})$ neighboring data values $z(\mathbf{u}_\alpha)$, $\alpha = 1, \ldots, n(\mathbf{u})$. If the estimate $z_{SK}^*(\mathbf{u})$ is restricted to be a linear combination of the data, it is written:

$$z_{\text{SK}}^*(\mathbf{u}) - m = \sum_{\alpha=1}^{n(\mathbf{u})} \lambda_\alpha^{SK}(\mathbf{u})\,[z(\mathbf{u}_\alpha) - m] = \boldsymbol{\lambda}^t \cdot \mathbf{D}, \qquad (3.17)$$

where $\boldsymbol{\lambda}$ is the column vector of the $n(\mathbf{u})$ kriging weights $\lambda_\alpha^{SK}(\mathbf{u})$ and $\mathbf{D}$ is the column vector of the $n(\mathbf{u})$ residual data values $[z(\mathbf{u}_\alpha) - m]$ built from the stationary and assumed known mean value m. Consideration of the residual variables rather than the original Z-variables ensures unbiasedness defined as zero expected error:

$$E\left\{Z_{\text{SK}}^*(\mathbf{u}) - Z(\mathbf{u})\right\} = 0. \qquad (3.18)$$

Remarks

- The zero expected error should be understood as a zero average error if the same geometric configuration of $n(\mathbf{u})$ data were applied elsewhere in the stationary zone, therefore in effect averaging over all possible combinations of the $n(\mathbf{u})$ data values. Ideally, one should ensure unbiasedness conditional to both data configuration and data values (the $z(\mathbf{u}_\alpha)$s), that is:

$$E\left\{Z_{\text{SK}}^*(\mathbf{u}) - Z(\mathbf{u}) | Z(\mathbf{u}_\alpha) = z(\mathbf{u}_\alpha), \alpha = 1, \ldots, n(\mathbf{u})\right\} = 0. \qquad (3.19)$$

Conditional unbiasedness for all possible combinations of data values $z(\mathbf{u}_\alpha)$ entails unbiasedness as in relation (3.18), not the reverse (Journel and Huijbregts, 1978; Goovaerts, 1997, p.182; Deutsch and Journel, 1998, p.94). SK, as most other linear estimators such as inverse distance-based, ensures unbiasedness, not conditional unbiasedness. This limitation of the unbiasedness property is at the source of many disappointments (David, 1977; Isaaks, 2005; Isaaks and Srivastava, 1989).

- A critical decision affecting the quality of estimate (3.17) is the choice of the $n(\mathbf{u})$ data retained to estimate any unsampled location $\mathbf{u}$. Consistency with the stationarity decision made to infer the covariance model would call for all locations $\mathbf{u} \in S$ to be estimated from the same data set (n) including all samples available over S. Such kriging with a "global" neighborhood is rarely implemented in practice, precisely because of defiance towards the decision of stationarity (Deutsch and Journel, 1998, p.32; Goovaerts, 1997, p.178). Not only must one decide the extent of the neighborhood within which the $n(\mathbf{u})$ data should be collected, but one may want to privilege certain directions, for example the direction of maximum continuity starting from $\mathbf{u}$ (Deutsch and Journel, 1998, p.33; Goovaerts, 1997, p.178).

Convexity issue

A definite advantage of kriging over traditional linear interpolators is that it is non-convex: the kriging estimate need not be valued in the interval of the data values retained (Goovaerts, 1997, p.177). For example, the SK estimate $z_{SK}^*(\mathbf{u})$ may be valued greater than the largest datum value $\max\{z(\mathbf{u}_\alpha), \alpha = 1, \ldots, n(\mathbf{u})\}$.

This advantage can turn into an inconvenience if the estimate $z_{SK}^*(\mathbf{u})$ is valued outside the z-physical bounds, for example a negative estimate for a positive variable such as a metal grade. One solution (not the best) to ensure convexity is to enforce the kriging weights to be all positive and sum up to 1; ordinary kriging weights do add up to 1 but are not necessarily all positive, see hereafter Section 3.6.2; and Barnes and Johnson (1984) and Rao and Journel (1996).

Simple kriging system

If the estimation criterion is chosen to be "least squared error," often a decision of mere convenience, the weights $\lambda_\alpha^{SK}(\mathbf{u})$ are given by a kriging system of linear equations built from a covariance model (Luenberger, 1969; Matheron, 1970; Journel and Huijbregts, 1978; Goovaerts, 1997, p.127):

$$\mathbf{K} \cdot \lambda = \mathbf{k} \qquad (3.20)$$

where $\mathbf{k}^T = [C(\mathbf{u} - \mathbf{u}_\alpha), \alpha = 1, \ldots, n(\mathbf{u})]$ is the data-to-unknown row covariance vector, $\mathbf{K} = [C(\mathbf{u}_\alpha - \mathbf{u}_\beta), \alpha, \beta = 1, \ldots, n(\mathbf{u})]$ is the data-to-data square covariance matrix; both matrices are built from the prior stationary covariance model:

$$C(\mathbf{h}) = \text{Cov}\{Z(\mathbf{u}), Z(\mathbf{u} + \mathbf{h})\} = C(0) - \gamma(\mathbf{h}), \qquad (3.21)$$

$C(0) = Var\{Z(\mathbf{u})\}$ is the stationary variance, and $2\gamma(\mathbf{h}) = Var\{Z(\mathbf{u}) - Z(\mathbf{u} + \mathbf{h})\}$ is the corresponding stationary variogram model.

The two main contributions of kriging to estimation are (Journel and Huijbregts, 1978) as follows.

1. The utilization of a variogram distance $\gamma(\mathbf{h})$ specific to the variable $Z(\mathbf{u})$ and the stationary zone S under study as opposed to a non-specific Euclidean distance $\mathbf{h}$, as used in inverse distance interpolation, for example. The variogram distance could be anisotropic, for example privileging data along the direction of greater continuity from the location $\mathbf{u}$ being estimated.
2. The consideration of the data-to-data covariance matrix $\mathbf{K}$ allows "data declustering", which leads to giving less weight to redundant data within a cluster of data as opposed to isolated data. This property of kriging allows correcting for bias due to preferential clustering of data, a common occurrence in earth sciences.

Kriging variance

A by-product of kriging and more generally of any least square regression is the estimation variance or kriging variance, which is the expected squared error whose minimization led to the kriging system (3.20) (Goovaerts, 1997, p.179; Chilès and Delfiner, 1999, p.156):

$$\sigma_{SK}^2(\mathbf{u}) = Var\{Z(\mathbf{u}) - Z_{SK}^*(\mathbf{u})\} = C(0) - \lambda^t \cdot \mathbf{k}. \qquad (3.22)$$

This variance is often misused as a measure of accuracy of the estimator $Z_{SK}^*(\mathbf{u})$. The variance expression (3.22) is data value-independent; it depends only on the spatial configuration of the data set $n(\mathbf{u})$ retained and the covariance model adopted; it is thus a covariance-dependent ranking index of data configuration, a valuable index for comparison of alternate data configurations but not yet a measure of estimation accuracy (Journel and Rossi, 1989). A better measure of the potential error associated with the estimator $Z_{SK}^*(\mathbf{u})$ would be the conditional error variance which is also dependent on the data values $z(\mathbf{u}_\alpha)$ (Goovaerts, 1997, p.180):

$$\sigma_{SK}^2(\mathbf{u}) = Var\{Z(\mathbf{u}) - Z_{SK}^*(\mathbf{u}) | Z(\mathbf{u}_\alpha) = z(\mathbf{u}_\alpha), \alpha = 1, \ldots, n(\mathbf{u})\}.$$

It can be shown that the kriging variance (3.22) is the average of the conditional error variance over all possible joint realizations of the $n(\mathbf{u})$ data values, the data configuration being fixed (Goovaerts, 1997, pp.180, 361). In the general case, one cannot ignore the impact of the actual data values on estimation accuracy.

One notable exception is that provided by a Gaussian RF model where *all* constitutive RVs $Z(\mathbf{u})$, $\mathbf{u} \in S$, are assumed jointly Gaussian-distributed. In that case, the previous conditional error variance is indeed data values-independent and identifies the kriging variance (3.22), a property known as homoscedasticity (Goovaerts, 1997, pp.82, 180; Anderson, 2003).

Distribution of error

Even if the kriging variance (3.22) could be retained as a representative error variance, one would know only two moments of the error distribution: the mean equal to zero per unbiasedness and the kriging variance. A two-parameter distribution would have to be assumed, for example a Gaussian error distribution. There is, unfortunately, little justification for a Gaussian error; the central limit theorem does not apply to spatial interpolation errors essentially because data and resulting errors are not independent one from another. If a Gaussian distribution assumption is accepted for convenience, one should be aware that this distribution decays rapidly at the tails (very small frequency of extreme values) and its adoption is very consequential, particularly when assessing the probability of large errors.

3.6.2 Ordinary kriging and other variants

The most restrictive aspect of any probabilistic approach is associated with the decision of stationarity (Section 3.6) which allows scanning a data set for replicates, averaging the latter to infer the required statistics. For example, the inference of the covariance model $C(\mathbf{h})$ needed to build any kriging system calls for pooling together pairs of data, $\{z(\mathbf{u}_\alpha), z(\mathbf{u}_\alpha + \mathbf{h})\}$, $\{z(\mathbf{u}_\beta), z(\mathbf{u}_\beta + \mathbf{h})\}$, approximately separated by the same vector $\mathbf{h}$ but otherwise taken at different locations $\mathbf{u}_\alpha$ and $\mathbf{u}_\beta$. Once that covariance model is available, the harsh consequences of the stationarity decision settle in. Rigorous theory would demand that the covariance model $C(\mathbf{h})$ and the corresponding stationary mean m be frozen over the area S of the stationarity decision. Yet in many applications, local information calls for a locally variable mean and sometimes for aspects of the covariance model to be made locally variable, for example the direction of anisotropy. Variants of the previous simple kriging system were developed to allow such flexibility, all amounting to a deviation from rigorous theory.

In ordinary kriging (OK) the expected value of the random function is locally re-estimated from local data, while the covariance model is kept stationary. The OK concept has been extended to local estimation of the parameters of a functional trend (KT or kriging with a trend). A locally varying mean (LVM) can also be input directly and used in expression (3.17) in lieu of the stationary mean m.

All these variants of kriging amount to relax the decision of stationarity initially necessary to define the random function model and infer its constitutive statistics, for example a variogram model. There have been many attempts at extending the RF theory to justify such liberty with regard to the original restrictive decision of stationarity (Matheron, 1970; Chilès and Delfiner, 1999, p.243; Goovaerts, 1997, p.143). The RF-originated tools can be modified to a reasonable extent with proper documentation. Possibly the best argument for such lax consideration of RF theory is realizing that there would be no practical geostatistics without OK or without kriging with a local varying mean and, more generally, without the code implementations that make it "work."

Ordinary kriging

The simple kriging (SK) expression (Eq. (3.17)) appears as an estimate of an unknown deviation from a known stationary mean m. If that mean is considered locally variable, it can be estimated from the same $n(\mathbf{u})$ local data used in expression (3.17); the corresponding estimate then takes its ordinary kriging (OK) expression (Goldberger, 1962; Matheron, 1970; Goovaerts, 1997, p.132; Journel and Huijbregts, 1978):

$$z_{OK}^*(\mathbf{u}) = \sum_{\alpha=1}^{n(\mathbf{u})} \lambda_\alpha^{OK}(\mathbf{u}) z(\mathbf{u}_\alpha), \qquad (3.23)$$

where the kriging weights sum to 1: $\sum_{\alpha=1}^{n(\mathbf{u})} \lambda_\alpha^{OK}(\mathbf{u}) = 1$.

The corresponding kriging system is similar to the SK system (Eq. (3.20)) with, in addition, one Lagrange parameter and equation to account for the above constraint on the OK weights. The qualifier "ordinary" is appropriate since OK is used more often than SK thanks to its robustness against local departures from the original decision of stationarity.

Kriging with a trend

A locally varying unknown mean $m(\mathbf{u})$ can be modeled by a function of the coordinates $\mathbf{u}$. That function, called a trend function, is of known shape or type but with unknown locally variable parameters; therefore the mean $m(\mathbf{u})$ remains unknown at any location $\mathbf{u}$. In the space domain where $\mathbf{u} = (x, y, z)$ coordinates, the trend function is usually a polynomial function of the coordinates; for example a trend linear in the horizontal space (x, y) but possibly quadratic in the vertical (z) would be written (Goovaerts, 1997, p.141; Deutsch and Journel, 1998, p.67):

$$E\{Z(\mathbf{u})\} = m(\mathbf{u}) = a_0(\mathbf{u}) + a_1(\mathbf{u})x + a_2(\mathbf{u})y + a_3(\mathbf{u})z + a_4(\mathbf{u})z^2. \qquad (3.24)$$

The five parameters $a_{\cdot}(\mathbf{u})$ are unknown and are estimated from the $n(\mathbf{u})$ local data available. The location coordinates $\mathbf{u} = (x, y, z)$ are known.

Similarly, in the time domain, a time series $Z(t)$ may present a periodic trend modeled as a cosine function of known frequency ω but unknown varying phase and amplitude $a_0(t)$, $a_1(t)$:

$$E\{Z(t)\} = m(t) = a_0(t) + a_1(t) \cdot \cos(2\pi\omega t). \qquad (3.25)$$

However, such cosine trend function is not yet programmed in SGeMS.

Once the unknown parameters $a_i(\mathbf{u})$ are estimated (implicitly and by a form of kriging), a simple kriging of type (3.17) is applied by replacing, at each location $\mathbf{u}$, the constant stationary mean m by the resulting mean estimate $m^*(\mathbf{u})$. The corresponding estimate is said to be a kriging with a trend model (KT) and is written:

$$z_{KT}^*(\mathbf{u}) = \sum_{\alpha=1}^{n(\mathbf{u})} \lambda_\alpha^{KT}(\mathbf{u})z(\mathbf{u}_\alpha). \qquad (3.26)$$

The KT weights $\lambda_\alpha^{KT}(\mathbf{u})$ are given by a kriging system similar to the SK system (Eq. (3.20)) but with additional constraints on the KT weights (Goldberger, 1962; Goovaerts, 1997, p.139; Journel and Huijbregts, 1978).

Actually OK is but a particular case of KT when the trend model (Eq. (3.24)) reduces to the sole term $a_0(\mathbf{u})$. In cases of interpolation with the $n(\mathbf{u})$ data surrounding the location $\mathbf{u}$ on both sides of the trend, OK would give results very close to KT. The specification of the trend functional type, say linear or quadratic, matters only in cases of extrapolation (Journel and Rossi, 1989; Goovaerts, 1997, p.147).

Kriging with a local varying mean

There are applications where some ancillary information (different from the z-data) provides at *all* locations the locally varying mean (LVM), then denoted $m^*(\mathbf{u})$. The SK expression (3.17) is then applied directly to the deviations from these locally varying mean values (Goovaerts, 1997, p.190):

$$z_{LVM}^*(\mathbf{u}) - m^*(\mathbf{u}) = \sum_{\alpha=1}^{n(\mathbf{u})} \lambda_\alpha^{SK}(\mathbf{u})\left[z(\mathbf{u}_\alpha) - m^*(\mathbf{u}_\alpha)\right]. \qquad (3.27)$$

Non-linear kriging

The qualifier "non-linear" applied to kriging is misleading if it promises some major breakaway from the fundamental limitation of considering only linear combinations of data. Most so-called non-linear kriging, including kriging of normal score transform as used in program *SGSIM* (Deutsch and Journel, 1998, p.75), disjunctive kriging (Matheron, 1973) or indicator kriging (Journel, 1983), are in fact all linear kriging applied to a non-linear transform of the variables involved. For example, lognormal kriging is but a kriging applied on the logarithms of the data (Rendu, 1979; Journel, 1980).

A non-linear transform of the original variable(s) is warranted if that transform allows defining new variables that:

- are more attuned to the problem addressed, as is the case for indicator transform (see hereafter Section 3.6.5);
- satisfy better the requirements of the algorithm being considered, e.g. the normal score transform allows meeting one requirement of the sequential Gaussian simulation algorithm (SGeMS code *SGSIM*, see Section 8.1.2 and Goovaerts, 1997, p.380), that the univariate distributions be Gaussian;
- exhibit better spatial correlation.

Remarks

Any non-linear back transform of kriging results may lead to severe biases if not carefully attended to. Non-robust unbiasedness corrections associated to back transform of kriging estimates may wipe out any benefit brought by working on the transformed variable. This is especially true with lognormal kriging: what is gained by working on the logarithm may be lost through a back transform that involves an exponentiation (Journel, 1980; Chilès and Delfiner, 1999, p.191; Deutsch and Journel, 1998, p.76). Note that indicator kriging estimates are used as probability estimates, without any back transform, see hereafter. Similarly, the kriging means and variances of normal score values are used directly to build the conditional distributions in sequential Gaussian simulation, these kriging results are never back transformed; it is the final simulated values that are back transformed, then indeed that back transform is sensitive to the tail extrapolation decision (Deutsch and Journel, 1998, p.135; Goovaerts, 1997, p.385).

A non-linear transform of the variable does not remove the most fundamental limitation of all kriging, which is that the data are related to the unknown one at a time; see the right-hand side covariance matrix **k** in the SK system (Eq. (3.20)). Kriging remains, however, the ultimate estimation paradigm if applied to functions of the data taken two by two, three by three, and ultimately taken altogether as a single multiple-point event, see Sections 3.6.5 and 3.7.

3.6.3 Kriging with linear average variable

An application of kriging with linear average variable is to estimate the average grade of a mining block from neighboring data which are both core grades and average grades of already mined-out blocks (Journel and Huijbregts, 1978; David, 1977). Another application is that related to tomographic imaging where data are defined as linear average over diverse ray paths (1D volume data) (Gómez-Hernández *et al.*, 2005; Hansen *et al.*, 2006). Kriging systems that use data at different volume support call for the covariance between any two block-support z-values, which happens to be a linear average of the point covariance model $C(\mathbf{h})$; see Journel and Huijbregts (1978). Any kriging system can be used with linear average data as long as the variogram/covariance values are properly regularized (averaged).

Covariance averaging

The covariance relating a point support value $Z(\mathbf{u})$ to a linear average $B_V(\mathbf{s})$ defined over a block of support volume V centered at $\mathbf{s}$ is derived from the point-to-point covariance model $C(\mathbf{u}, \mathbf{u} + \mathbf{h}) = C(\mathbf{h})$ as:

$$\bar{C}(\mathbf{u}, V(\mathbf{s})) = \mathrm{Cov}\{B_V(\mathbf{s}), Z(\mathbf{u})\} = \frac{1}{|V|} \int_{\mathbf{u}' \in V(\mathbf{s})} C(\mathbf{u} - \mathbf{u}') d\mathbf{u}'.$$

Similarly, an average block-to-block covariance is given as:

$$\bar{C}(V, V') = \int_{\mathbf{u} \in V} \int_{\mathbf{u}' \in V'} C(\mathbf{u} - \mathbf{u}') d\mathbf{u} d\mathbf{u}'.$$

These upscaled or regularized covariances provide a valid model of point-to-block and block-to-block covariances; they are used whenever linear average data are present. Fast calculation of each block-averaged variogram/covariance is discussed in Section 7.4 and Kyriakidis *et al.* (2005).

Kriging with block and point data

Data taken at different scales, both on block-support and on point-support, can be considered simultaneously in the kriging system. The only condition is that all block data are linear averages of point values. For simplicity, the kriging theory is illustrated here with simple kriging.

The block data $B(\mathbf{v}_\alpha)$ is defined as the spatial linear average of point values $Z(\mathbf{u}')$ within the block volume $\mathbf{v}_\alpha$ (Journel and Huijbregts, 1978; Hansen *et al.*, 2006; Liu *et al.*, 2006b; Goovaerts, 1997, p.152):

$$B(\mathbf{v}_\alpha) = \frac{1}{|\mathbf{v}_\alpha|} \int_{\mathbf{v}_\alpha} L_\alpha(Z(\mathbf{u}')) d\mathbf{u}' \quad \forall \alpha \tag{3.28}$$

where L_α is a known linear averaging function.

The simple kriging estimator $Z^*_{\mathrm{SK}}(\mathbf{u})$ conditioned to both point and block data is then written:

$$Z^*_{\mathrm{SK}}(\mathbf{u}) - m = \Lambda^t \cdot \mathbf{D} = \sum_{\alpha=1}^{n(\mathbf{u})} \lambda_\alpha(\mathbf{u}) \cdot [D(\mathbf{u}_\alpha) - m] \tag{3.29}$$

where $\Lambda^t = [\lambda_P \ \lambda_B]$ denotes the kriging weights for point data P and block data B; $\mathbf{D}^t = [\mathbf{P} \ \mathbf{B}]$ denotes the data value vector; $D(\mathbf{u}_\alpha)$ is a specific datum at location $\mathbf{u}_\alpha$; $n(\mathbf{u})$ denotes the number of data; and m denotes the stationary mean.

The kriging weights Λ are obtained through the kriging system:

$$\mathbf{K} \cdot \Lambda = \mathbf{k} \tag{3.30}$$

with

$$\mathbf{K} = \begin{bmatrix} \mathbf{C}_{PP} & \bar{\mathbf{C}}_{PB} \\ \bar{\mathbf{C}}_{PB}^T & \bar{\mathbf{C}}_{BB} \end{bmatrix} \text{ and } \mathbf{k} = \begin{bmatrix} \mathbf{C}_{PP_0} \\ \bar{\mathbf{C}}_{BP_0} \end{bmatrix}$$

where $\mathbf{K}$ denotes the data-to-data covariance matrix, $\mathbf{k}$ denotes the data-to-unknown covariance matrix, $\mathbf{C}$ denotes the point covariance submatrix, $\bar{\mathbf{C}}$ denotes a covariance submatrix involving a block support and P_0 is the estimation location.

The kriging variance is then written:

$$\sigma_{\text{SK}}^2(\mathbf{u}) = \text{Var}\left\{Z(\mathbf{u}) - Z_{\text{SK}}^*(\mathbf{u})\right\} = C(0) - \Lambda^t \cdot \mathbf{k}$$

with $C(0) = \text{Var}\{Z(\mathbf{u})\}$ being the stationary variance.

3.6.4 Cokriging

There is nothing in the theory of kriging and expressions (3.17) to (3.27) that constrains the unknown $Z(\mathbf{u})$ and the data $Z(\mathbf{u}_\alpha)$ to relate to the same attribute. One can extend the notation Z to different attribute values $Z_k(\mathbf{u})$, $Z_{k'}(\mathbf{u}_\alpha)$, say for estimation of a porosity value $Z_k(\mathbf{u})$ at location $\mathbf{u}$ from porosity data $Z_k(\mathbf{u}_\alpha)$ and seismic amplitude data $Z_{k'}(\mathbf{u}_\alpha)$ with $k' \neq k$ at neighboring locations $\mathbf{u}_\alpha$. Cokriging is the extension of the kriging paradigm to estimation of one attribute using data related to other attributes (Myers, 1982; Wackernagel, 1995; Goovaerts, 1997, p.203; Chilès and Delfiner, 1999, p.296).

For example, the simple cokriging estimate of an unsampled porosity $z_1(\mathbf{u})$ from $n_1(\mathbf{u})$ neighboring porosity data $z_1(\mathbf{u}_\alpha)$ and $n_2(\mathbf{u})$ seismic data $z_2(\mathbf{u}_\beta')$ would be written:

$$z_1^*(\mathbf{u}) - m_1 = \sum_{\alpha=1}^{n_1(\mathbf{u})} \lambda_\alpha \left[z_1(\mathbf{u}_\alpha) - m_1\right] + \sum_{\beta=1}^{n_2(\mathbf{u})} \lambda_\beta \left[z_2(\mathbf{u}_\beta') - m_2\right], \tag{3.31}$$

where m_1 and m_2 are the two stationary means.

The only difficulty, but a serious one in practice, comes from the necessity to infer and model *jointly* many cross-covariance/variogram models, up to K^2 models (K is the total number of attributes) in the case of cross-covariances. If it is already difficult to infer the variogram of a single variable in a 3D possibly anisotropic space, real data are rarely enough to infer a set of cross-variograms for more than $K = 3$ different attribute variables.

In order to alleviate the burden of modeling all these variograms with the linear model of coregionalization (Goovaerts, 1997, p.108; Chilès and Delfiner, 1999, p.339), various shortcut models have been proposed, two of which are available in SGeMS: the Markov Model 1 and the Markov Model 2 (Almeida and Journel, 1994; Journel, 1999; Rivoirard, 2004; Chilès and Delfiner, 1999, p.305; Goovaerts, 1997, p.237). The Markov screening hypothesis also entails that only the secondary data coincidental with the estimation location need to be retained for the estimation. This restricts the size of the kriging matrix to $n + 1$, where n is the number of hard conditioning data in the vicinity of the estimation location. For text clarity, we will only consider the case of a single secondary variable ($K = 2$).

Markov Model 1 The Markov Model 1 (MM1) considers the following Markov-type screening hypothesis:

$$E\{Z_2(\mathbf{u})|Z_1(\mathbf{u}); Z_1(\mathbf{u}+\mathbf{h})\} = E\{Z_2(\mathbf{u})|Z_1(\mathbf{u})\}$$

i.e. the dependence of the secondary variable on the primary is limited to the co-located primary variable. The cross-covariance is then proportional to the auto-covariance of the primary variable:

$$C_{12}(\mathbf{h}) = \frac{C_{12}(0)}{C_{11}(0)}C_{11}(\mathbf{h}) \tag{3.32}$$

where C_{12} is the cross-covariance between the two variables Z_1 and Z_2 and C_{11} is the covariance of the primary variable Z_1. Solving the cokriging system under the MM1 model only calls for knowledge of C_{11}, hence the inference and modeling effort is the same as for kriging with only primary Z_1-data. Although very congenial the MM1 model should not be used when the support of the secondary variable Z_2 is larger than the one of Z_1, lest the variance of Z_1 would be underestimated. It is better to use the Markov Model 2 in that case.

Markov Model 2 The Markov Model 2 (MM2) was developed for the case where the volume support of the secondary variable is larger than that of the primary variable (Journel, 1999). This is often the case with remote sensing and seismic-related data. The more relevant Markov-type hypothesis is now:

$$E\{Z_1(\mathbf{u})|Z_2(\mathbf{u}); Z_2(\mathbf{u}+\mathbf{h})\} = E\{Z_1(\mathbf{u})|Z_2(\mathbf{u})\}$$

i.e. the dependence of the primary variable on the secondary is limited to the co-located secondary variable. The cross-variogram is now proportional to the covariance of the secondary variable:

$$C_{12}(\mathbf{h}) = \frac{C_{12}(0)}{C_{11}(0)}C_{22}(\mathbf{h}). \tag{3.33}$$

In order for all three covariances C_{11}, C_{12} and C_{22} to be consistent, C_{11} is modeled as a linear combination of C_{22} and any permissible residual correlation ρ_R. Expressed in terms of correlograms

$$\rho_{11}(\mathbf{h}) = \frac{C_{11}(\mathbf{h})}{C_{11}(0)}, \quad \rho_{22}(\mathbf{h}) = \frac{C_{22}(\mathbf{h})}{C_{22}(0)}$$

which is written as:

$$\rho_{11}(\mathbf{h}) = \rho_{12}^2 \cdot \rho_{22}(\mathbf{h}) + (1 - \rho_{12}^2)\rho_R(\mathbf{h}) \tag{3.34}$$

where ρ_{12} is the co-located coefficient of correlation between $Z_1(\mathbf{u})$ and $Z_2(\mathbf{u})$.

Independent of the cross-covariance modeling scheme adopted, cokriging shares all the contributions and limitations of kriging: it provides a linear, least squared error, regression combining data of diverse types accounting for their redundancy and respective variogram distances to the unknown. Cokriging considers the data one at a time and the cokriging variance being data value-independent is an incomplete measure of estimation accuracy. The linear limitation of kriging may be here more serious, since cokriging would ignore any non-linear relation between two different attributes which could be otherwise capitalized upon for cross-estimation. One possible solution is to apply cokriging on non-linear transforms of the original variables.

The kriging system with block data as described in Section 3.6.3 can also be seen as a cokriging system, where the cross-dependence between points and block is given by the regularization process.

3.6.5 *Indicator kriging*

Indicator kriging is yet another form of kriging but applied to variables that are binary indicators of occurrence of an event:

$$I_k(\mathbf{u}) = \begin{cases} 1 & \text{if the event } k \text{ occurs at location } \mathbf{u} \\ 0 & \text{if not} \end{cases} \tag{3.35}$$

or for the continuous case:

$$I(\mathbf{u}; z_k) = \begin{cases} 1 & \text{if } Z(\mathbf{u}) \le z_k \\ 0 & \text{if not} \end{cases}.$$

The event k to be estimated could be presence of facies of type k at location $\mathbf{u}$, or could be that the unsampled continuous variable $Z(\mathbf{u})$ is valued below the threshold z_k.

The particularity of indicator kriging (IK) is that it delivers a kriging estimate that can be interpreted directly (without any transform) as an estimate of the probability for the unsampled event to occur at location $\mathbf{u}$ conditional to the observed data set $n(\mathbf{u})$ (Goovaerts, 1997, p.293; Chilès and Delfiner, 1999, p.383; Journel, 1983). The IK estimate is hereafter written under its simple kriging form:

$$I_{SK}^*(\mathbf{u}) = \text{Prob}^* \{I(\mathbf{u}) = 1 | n(\mathbf{u})\}$$

$$= \sum_{\alpha=1}^{n(\mathbf{u})} \lambda_\alpha(\mathbf{u}) I_k(\mathbf{u}_\alpha) + \left[1 - \sum_{\alpha=1}^{n(\mathbf{u})} \lambda_\alpha(\mathbf{u}) \right] \cdot p_0, \qquad (3.36)$$

where $p_0 = E\{I(\mathbf{u})\} = \text{Prob}\{I(\mathbf{u}) = 1\}$ is the prior probability of the event occurring, $\lambda_\alpha(\mathbf{u})$ is the kriging weight associated to the indicator datum $I(\mathbf{u}_\alpha)$ valued 0 or 1.

If soft information is available providing a location-specific prior probability $p(\mathbf{u})$, that probability could replace p_0 in expression (3.36). Indicator kriging can be seen as an updating of that prior probability $p(\mathbf{u})$ by the indicator data $I(\mathbf{u}_\alpha)$ (Goovaerts, 1997, p.293).

The fact that kriging is not a convex estimator valued between the minimum and maximum indicator data value (here 0 and 1) is particularly damaging in that IK may provide estimated probabilities outside the interval $[0, 1]$. Order relation corrections are then implemented. An alternative is to consider multiple-point (mp) statistics through products instead of linear combinations of the indicator data, see the following presentation on extended normal equations and in Section 3.10 the nu/tau model (Journel, 2002).

On the positive side, kriging being an exact estimator, if estimation is performed at any hard datum location $\mathbf{u}_\alpha$, the resulting probability estimate is "hard", that is valued 0 or 1 identifying that hard datum value. If the indicator variogram used is continuous with a small nugget effect the probability estimate would smoothly depart from that hard value (0 or 1) as the location $\mathbf{u}$ to be estimated gets away from $\mathbf{u}_\alpha$.

3.7 An introduction to mp statistics

Consider again the linear indicator kriging expression (3.36). First note that any non-linear transform of a binary (indicator) variable is non-effective in that it results in just another binary variable. To extract more from the indicator data set $\{I(\mathbf{u}_\alpha), \alpha = 1, \ldots, n(\mathbf{u})\}$, one needs to consider these data two by two, three by three, $\ldots$, at the limit altogether as a single data event.

Consider then the extended IK expression (3.37) which is a linear combination:

- of the indicator data taken one at a time as in expression (3.36), there are $n(\mathbf{u})$ such indicator data;
- of the indicator data taken two at a time; there are a number $\binom{n(\mathbf{u})}{2}$ of combinations of such pairs;

- of the indicator data taken three at a time; there are $\binom{n(\mathbf{u})}{3}$ such triplets;
- ...
- of the indicator data taken altogether; there is only one such product:

$$I_{SK}^*(\mathbf{u}) = \text{Prob}^* \{I(\mathbf{u}) = 1 | n(\mathbf{u})\}$$

$$= p_0 \quad \text{(prior probability for } I(\mathbf{u}) = 1)$$

$$+ \sum_{\alpha=1}^{n(\mathbf{u})} \lambda_\alpha^{(1)}(\mathbf{u}) \left[I(\mathbf{u}_\alpha) - p_0\right] \quad \text{(one at a time)}$$

$$+ \sum_{\alpha=1}^{(n(\mathbf{u}),2)} \lambda_\alpha^{(2)}(\mathbf{u}) \left[I(\mathbf{u}_{\alpha_1})I(\mathbf{u}_{\alpha_2}) - E\{I(\mathbf{u}_{\alpha_1})I(\mathbf{u}_{\alpha_2})\}\right] \quad \text{(two at a time)}$$

$$+ \sum_{\alpha=1}^{(n(\mathbf{u}),3)} \lambda_\alpha^{(3)}(\mathbf{u}) \left[I(\mathbf{u}_{\alpha_1})I(\mathbf{u}_{\alpha_2})I(\mathbf{u}_{\alpha_3}) - E\{I(\mathbf{u}_{\alpha_1})I(\mathbf{u}_{\alpha_2})I(\mathbf{u}_{\alpha_3})\}\right]$$

$$\text{(three at a time)}$$

$$+ \cdots$$

$$+ \lambda_\alpha^{(n(\mathbf{u}))}(\mathbf{u}) \left[\prod_\alpha^{n(\mathbf{u})} I(\mathbf{u}_\alpha) - E\left\{\prod_\alpha^{n(\mathbf{u})} I(\mathbf{u}_\alpha)\right\}\right] \quad \text{(taken altogether).}$$

$$(3.37)$$

Remarks

- Expression (3.37) is a simple indicator (co)kriging estimator extended to include data taken 2, 3, up to all at a time. The corresponding simple kriging system is called "the extended system of normal equations" (Journel and Alabert, 1989); it has $2^{n(\mathbf{u})}$ equations yielding the $2^{n(\mathbf{u})}$ kriging weights $\lambda_\alpha^{(\cdot)}$. Expression (3.37) already includes the non-bias equation providing the weight given to the data p_0.
- Note that a set of $n(\mathbf{u})$ binary indicator data can take $2^{n(\mathbf{u})}$ possible joint outcomes, a number precisely equal to the number of kriging weights in the extended IK expression (3.37); indeed:

$$\sum_{\alpha=1}^{n(\mathbf{u})} \binom{n(\mathbf{u})}{\alpha} = 2^{n(\mathbf{u})}.$$

It can be shown that the solution of the full extended normal system (3.37) provides the exact conditional probability value for $I(\mathbf{u}) = 1$ for all possible data value combinations; there are $2^{n(\mathbf{u})}$ such combinations.

- The information carried by the product $I(\mathbf{u}_\alpha)I(\mathbf{u}_\beta)$ is not redundant with that carried by the two individual data $I(\mathbf{u}_\alpha)$ and $I(\mathbf{u}_\beta)$ or any of their linear combinations. By restricting expression (3.37) to indicator data taken one at a time as in expression (3.36), one is losing precious information provided by observation of the data taken jointly. The information $I(\mathbf{u}_\alpha)I(\mathbf{u}_\beta)$ sits there ready for the taking as a covariate through a form of cokriging. As in any other cokriging, it suffices to evaluate the additional covariances required which are, in the case of data taken only up to two at a time:

- the 3-point covariance linking any 2-point data $I(\mathbf{u}_\alpha)I(\mathbf{u}_\beta)$ to the unknown $I(\mathbf{u})$;
- the 4-point covariance linking any two doublets of data, say $I(\mathbf{u}_1)I(\mathbf{u}_2)$ with $I(\mathbf{u}_3)I(\mathbf{u}_4)$ and measuring the redundancy between these two doublets.

The traditional IK estimator (Eq. (3.36)) using the data one at a time calls only for traditional 2-point covariances. An extended IK estimator using data taken two at a time would require in addition 3- and 4-point covariances. An extended IK estimator using all possible combinations of data up to only two at a time would call for a kriging system of dimension $n(\mathbf{u}) + \binom{n(\mathbf{u})}{2}$; for example, if $n(\mathbf{u}) = 10$, then $n(\mathbf{u}) + \binom{n(\mathbf{u})}{2} = 10 + 45 = 55$, a considerable increase in dimension and covariance modeling effort! Clearly this is not practical, particularly if data are to be taken many more than two at a time.

The solution is to consider all $n(\mathbf{u})$ data grouped together into a single multiple-point data event DEV which relates to the last term of expression (3.37). The corresponding IK estimate is then written:

$$I_{SK}^*(\mathbf{u}) - p_0 = \lambda \cdot \left[DEV - E\{DEV\}\right]$$

with

$$DEV = \prod_\alpha^{n(\mathbf{u})} I(\mathbf{u}_\alpha; i_\alpha), \text{ with: } I(\mathbf{u}_\alpha; i_\alpha) = \begin{cases} 1 & \text{if } I(\mathbf{u}_\alpha) = i_\alpha \\ 0 & \text{if not.} \end{cases}$$

Note that the multiple-point random variable DEV is also binary equal to 1 *if and only if* all $n(\mathbf{u})$ indicator RVs $I(\mathbf{u}_\alpha)$ identify the indicator data i_α actually observed.

The corresponding kriging system reduces to one single equation, also called single normal equation, delivering the single data values-dependent weight λ. It can be shown that this single normal equation identifies the *exact* expression of the conditional probability:

$$
\begin{aligned}
I_{SK}^*(\mathbf{u}) &\equiv \text{Prob}\{I(\mathbf{u}) = 1 | n(\mathbf{u})\} \\
&= \frac{\text{Prob}\{I(\mathbf{u}) = 1, n(\mathbf{u})\}}{\text{Prob}\{n(\mathbf{u})\}} \\
&= \frac{\text{Prob}\{I(\mathbf{u}) = 1, I(\mathbf{u}_\alpha) = i_\alpha, \alpha = 1, \dots, n(\mathbf{u})\}}{\text{Prob}\{I(\mathbf{u}_\alpha) = i_\alpha, \alpha = 1, \dots, n(\mathbf{u})\}}.
\end{aligned}
\tag{3.38}
$$

Note that the probability in the numerator of the exact expression (3.38) is actually an $(n(\mathbf{u}) + 1)$-point covariance while the denominator is an $n(\mathbf{u})$-point covariance, both non-centered. Indeed, and as an example for the case of the specific two data values $I(\mathbf{u}_1) = 1$ and $I(\mathbf{u}_2) = 0$, that numerator is written as the 3-point non-centered covariance:

$$\text{Prob}\{I(\mathbf{u}) = 1, I(\mathbf{u}_1) = 1, I(\mathbf{u}_2) = 0\} = E\{I(\mathbf{u}) \cdot I(\mathbf{u}_1) \cdot [1 - I(\mathbf{u}_2)]\}.$$

Indicator kriging, when considering the $n(\mathbf{u})$ data altogether as a single multiple-point data event, identifies Bayes relation (Eq. (3.38)). Simulating the value $I(\mathbf{u})$ from probabilities of type (3.38) is at the root of the Single Normal Equation SIMulation algorithm *SNESIM* (Strebelle, 2000). Instead of modeling the mp covariance as a function of Euclidean distances $\mathbf{h}$ as one would model a 2-point covariance $C(\mathbf{h})$, the two mp covariance values appearing as numerator and denominator of expression (3.38) are lifted directly from a training image. In even simpler terms, the conditional probability $\text{Prob}\{I(\mathbf{u}) = 1|n(\mathbf{u})\}$ (Eq. (3.38)) is identified to the experimental proportion of those training replicates of the mp data events that feature at their center location $\mathbf{u}$ an event $I(\mathbf{u}) = 1$ (Strebelle, 2002).

In essence, the training image provides all the necessary mp covariance values; the decision of stationarity allows scanning a specific training image for replicates (exact or approximate) of the single mp conditioning data event. This is no different from scanning a training image for replicates of pairs of values allowing the modeling of a 2-point covariance/variogram. One may argue that with a variogram one borrows less from that training image; one then forgets that the missing mp stats are then implied in a non-controllable fashion by the simulation algorithm retained; indeed there cannot be any stochastic simulation without a full mp model; recall the previous discussion on the necessity of an mp model in Section 3.3.2. The better answer is that one does trust the training image characteristic structures and mp patterns and wishes to use them in the estimation/simulation exercise, and this cannot be done through 2-point statistics such as the variogram.

3.8 Two-point simulation algorithms

Traditional (2-point) simulation algorithms aim at reproducing a prior covariance $C(\mathbf{h})$ model, or equivalently a variogram model, that is a statistical relation between any two values $z(\mathbf{u})$ and $z(\mathbf{u} + \mathbf{h})$ in space. The missing information about what should be the relation in space of three or more values taken jointly is then *necessarily* provided by the simulation algorithm retained. Multiple-point structures imposed implicitly by the algorithm are most likely of high entropy nature, i.e. minimizing organization (Journel and Zhang, 2006).

If you wish the simulated realizations to reflect specific structures and patterns beyond 2-point correlation, these structures must be specified as input to a simulation algorithm that can reproduce them. Specific structures never occur by chance.

The covariance-based simulation algorithms widely used in practice stem from essentially two classes; the first class is anchored on the properties of the multivariate Gaussian RF model (Goovaerts, 1997, p.380), the second class builds on the interpretation of an indicator expected value as a conditional probability (Goovaerts, 1997, p.393), recall expression (3.36).

This initial release of SGeMS proposes the following well established covariance-based (2-point) simulation algorithms:

- *LUSIM*, or Gaussian simulation with LU decomposition, see Section 8.1.1 and Deutsch and Journel (1998, p.169),
- *SGSIM*, or sequential Gaussian simulation, see Section 8.1.2 and Deutsch and Journel (1998, p.170),
- *COSGSIM*, or sequential Gaussian co-simulation, see Section 8.1.3,
- *DSSIM*, or direct sequential simulation, see Section 8.1.4,
- *SISIM*, or sequential indicator simulation, see Section 8.1.5 and Deutsch and Journel (1998, p.175),
- *COSISIM*, or sequential indicator co-simulation, see Section 8.1.6,
- *BSSIM*, or block sequential simulation, see Section 8.1.7,
- *BESIM*, or block error simulation, see Section 8.1.8.

3.8.1 Sequential Gaussian simulation

The remarkably convenient properties of the Gaussian RF model explain its success, a quasi monopoly of probabilistic models for continuous variables. Indeed a Gaussian RF is fully characterized by its mean vector and covariance matrix; all conditional distributions are Gaussian, fully characterized by only two moments, the conditional mean and variance themselves given by simple kriging (Journel and Huijbregts, 1978; Anderson, 2003). Thus a Gaussian RF model would appear as the ultimate model when only 2-point statistics can be inferred. Unfortunately, the qualifier is that a Gaussian RF maximizes entropy (disorder) beyond the input covariance model (Chilès and Delfiner, 1999, p.412; Journel and Deutsch, 1993), hence a Gaussian-based simulation algorithm such as *SGSIM* cannot deliver any image with definite patterns or structures involving more than two locations at a time. The previous limitation matters little if one is simulating a "homogeneously heterogeneous" spatial distribution such as porosity or metal grade within the pre-defined geometry of a relatively homogeneous lithofacies or rock type.

In the *SGSIM* algorithm (Journel, 1993; Goovaerts, 1997, p.380) the mean and variance of the Gaussian distribution at any location along the simulation path is estimated by the kriging estimate and the kriging variance. The value drawn from that distribution is then used as conditioning data. Transform of the original data into a Gaussian distribution may be necessary and is normally performed by the normal score transform, see Section 8.1.2 for the SGeMS implementation.

3.8.2 Direct sequential simulation

It can be shown that reproduction of a covariance model does not require a Gaussian RF, but just that the mean and variance of every conditional distribution be those given by SK; the conditional distribution type need not be Gaussian, it can also vary from one simulation node to another (Journel, 1994; Bourgault, 1997). Consequently there is no need for any normal score transform and back transform. The sequential simulation can be performed *directly* with the original z-variable and data, hence the name "direct sequential simulation" (program *DSSIM*).

One main advantage of *DSSIM* is that the simulation can be made conditional to local linear average z-data. Indeed kriging can accommodate data defined on volume/block support as long as these data are linear average of z-values; see Section 3.6.3. The normal score transform being non-linear would undo such linearity. The absence of a prior transformation of the data in *DSSIM* makes it an algorithm of choice for "downscaling," a process whereby large scale block-support data are "un-averaged" into realizations of smaller support values (Kyriakidis and Yoo, 2005; Boucher and Kyriakidis, 2006). The *DSSIM* simulated values reproduce the target covariance model and honor whatever small support data are available; in addition their block averages match the corresponding block data (Hansen *et al.*, 2006).

The price to pay for the absence of normal score transform is absence of a back transform, hence there is no guarantee for the *DSSIM* simulated z-realizations to reproduce the z-data histogram.

Such global histogram reproduction can be obtained in two ways.

- A post-processing similar to the normal score back-transform done in Gaussian simulation. Such back-transform should be such as not to undo the data reproduction (Deutsch, 1996; Journel and Xu, 1994). The utility program *TRANS* discussed in Section 9.1 allows such transform honoring the original point-support data values; this, however, degrades the covariance reproduction.
- Put to use the large degree of freedom represented by the ability to choose at any simulation node any distribution type (Bourgault, 1997). The procedure retained by the *DSSIM* code consists of sampling only that part of a translated z-target histogram that matches the local SK mean and variance (Soares, 2001; Oz *et al.*, 2003).

3.8.3 Direct error simulation

In all generality one can express any unsampled value $z(\mathbf{u})$ as the sum of its estimate $z^*(\mathbf{u})$ plus the corresponding estimation error $r(\mathbf{u})$:

$$z(\mathbf{u}) = z^*(\mathbf{u}) + r(\mathbf{u}).$$

The estimated value $z^*(\mathbf{u})$ is known, but the error is not. Thus simulation would consist of simulating that error $r(\mathbf{u})$ under various constraints. For example, the simulated error should have mean zero and variance equal to the known kriging variance if $z^*(\mathbf{u})$ is obtained by kriging. As for the distribution from which the simulated error should be drawn, it may or may not be Gaussian. If the random variable error $R(\mathbf{u})$ is orthogonal (uncorrelated) to the random variable estimator $Z^*(\mathbf{u})$, as guaranteed by kriging (Luenberger, 1969; Journel and Huijbregts, 1978; Chilès and Delfiner, 1999, p.465), then the error value $r_s(\mathbf{u})$ can be drawn independently of the estimated value $z^*(\mathbf{u})$:

$$z_{cs}(\mathbf{u}) = z_K^*(\mathbf{u}) + r_s(\mathbf{u}) \qquad (3.39)$$

where: $z_K^*(\mathbf{u})$ is the kriging estimate;
    $r_s(\mathbf{u})$ is an error value drawn from a distribution with zero mean and variance equal to the kriging variance $\sigma_K^2(\mathbf{u}) = Var\left\{Z(\mathbf{u}) - Z_K^*(\mathbf{u})\right\}$;
    $z_{cs}(\mathbf{u})$ is the simulated value.

The simulated field $\{z_{cs}(\mathbf{u}), \mathbf{u} \in \text{study area}\}$

– honors the data value $z(\mathbf{u}_\alpha)$ at data location $\mathbf{u}_\alpha$ since $z_K^*(\mathbf{u}_\alpha) = z(\mathbf{u}_\alpha)$ per kriging exactitude;
– has the correct variance since:

$$\text{Var}\{Z_{cs}(\mathbf{u})\} = \text{Var}\left\{Z_K^*(\mathbf{u})\right\} + \text{Var}\{R(\mathbf{u})\}$$
$$= \text{Var}\left\{Z_K^*(\mathbf{u})\right\} + \left[\sigma_K^2(\mathbf{u}) = Var\left\{Z(\mathbf{u}) - Z_K^*(\mathbf{u})\right\}\right]$$

per orthogonality of the error $R(\mathbf{u})$ with $Z_K^*(\mathbf{u})$.

However, there remains to ensure that the simulated field $Z_{cs}(\mathbf{u})$ features the same covariance as $Z(\mathbf{u})$. This is obtained in sequential simulation (Sections 3.8.1 and 3.8.2) by adding into the kriging data set for $z_K^*(\mathbf{u})$ all previously simulated value $z_{cs}(\mathbf{u}')$ found in its neighborhood. An alternative is to simulate the error $r_s(\mathbf{u})$ by lifting it from an error training image sharing the same (non-stationary) covariance as the actual error $R(\mathbf{u}) = Z(\mathbf{u}) - Z_K^*(\mathbf{u})$. That error training image can be generated by repeating the estimation procedure used to generate $z_K^*(\mathbf{u})$ from data $z(\mathbf{u}_\alpha), \alpha = 1, \ldots, n$ on a non-conditional simulated realization $z_s(\mathbf{u})$ of the random function $Z(\mathbf{u})$ using the same geometric configuration of "simulated" data $z_s(\mathbf{u}_\alpha)$, $\alpha = 1, \ldots, n$. This process is written (Journel and Huijbregts, 1978; Deutsch and Journel, 1998, p.127; Chilès and Delfiner, 1999, p.465):

$$z_{\mathrm{cs}}^{(l)}(\mathbf{u}) = z_K^*(\mathbf{u}) + \left[z_s^{(l)}(\mathbf{u}) - z_{Ks}^{*(l)}(\mathbf{u}) \right] \tag{3.40}$$

where:

- $z_s^{(l)}(\mathbf{u})$ is the lth non-conditional simulated realization of the random field $Z(\mathbf{u})$ honoring its covariance model;
- $z_K^*(\mathbf{u})$ is the kriging estimate built from the actual data values $z(\mathbf{u}_\alpha)$, $\alpha = 1, \ldots, n$;
- $z_{Ks}^{*(l)}(\mathbf{u})$ is the kriging estimate built from the simulated data values $z_s^{(l)}(\mathbf{u}_\alpha)$ taken from the non-conditionally simulated field $z_s(\mathbf{u})$ at the actual data locations $\mathbf{u}_\alpha, \alpha = 1, \ldots, n$;
- $z_{\mathrm{cs}}^{(l)}(\mathbf{u})$ is the lth conditionally simulated realization.

Warning: a histogram input is required to simulate the intermediate unconditional realizations, however that histogram may not be reproduced in the final simulated conditional realizations.

Note that the kriging weights $\lambda_\alpha(\mathbf{u})$ used for both kriging estimates $z_K^*(\mathbf{u})$ and $z_{Ks}^{*(l)}(\mathbf{u})$ are the same, since the simulated field $Z_s(\mathbf{u})$ shares the same covariance model and the same data geometry as the actual field $Z(\mathbf{u})$. There lies the main (potential) advantage of the direct error simulation approach: one single kriging is needed per simulation node $\mathbf{u}$ no matter the number L of conditional simulated realizations $z_{\mathrm{cs}}^{(l)}(\mathbf{u})$, $l = 1, \ldots, L$, needed. One could then utilize any fast non-conditional simulation algorithm to generate the L required fields $z_s^{(l)}(\mathbf{u})$ (Chilès and Delfiner, 1999, pp.494, 513; Oliver, 1995; Lantuéjoul, 2002); the L krigings to obtain the $z_{Ks}^{*(l)}(\mathbf{u})$ are then obtained very fast by mere matrix multiplication from the stored kriging weights $\lambda_\alpha(\mathbf{u})$; last, an addition (Eq. (3.39)) gives the L conditional fields $z_{\mathrm{cs}}^{(l)}(\mathbf{u})$.

The caveat though is that the error field $R(\mathbf{u}) = Z(\mathbf{u}) - Z_K^*(\mathbf{u})$ must be independent (or at least uncorrelated) to the estimated signal $Z_K^*(\mathbf{u})$. This is not a trivial requirement, only guaranteed if simple kriging is applied to a multiGaussian field $Z(\mathbf{u})$.

3.8.4 Indicator simulation

Indicator simulation was introduced to simulate categorical variables defined through a set of K binary indicator variables (Journel, 1983; Goovaerts, 1997, p.423; Chilès and Delfiner, 1999, p.512). The algorithm was later extended to simulation of a continuous variable made discrete over K classes. Consider the corresponding two definitions:

$$I_k(\mathbf{u}) = \begin{cases} 1 & \text{if the category } k \text{ occurs at } \mathbf{u} \\ 0 & \text{if not} \end{cases} \tag{3.41}$$

or

$$I(\mathbf{u}; z_k) = \begin{cases} 1 & \text{if } Z(\mathbf{u}) \le z_k \\ 0 & \text{if not} \end{cases}.$$

Indicator kriging (Section 3.6.5) would provide estimates of the K class probabilities conditional to the local data set $n(\mathbf{u})$, respectively:

$$\text{Prob}\{\mathbf{u} \in k | n(\mathbf{u})\} \in [0, 1], \quad k = 1, \ldots, K, \tag{3.42}$$

with: $\sum_{k=1}^{K} \text{Prob}\{\mathbf{u} \in k | n(\mathbf{u})\} = 1$, or

$$\text{Prob}\{Z(\mathbf{u}) \le z_k | n(\mathbf{u})\} \in [0, 1], \quad k = 1, \ldots, K,$$

with $\text{Prob}\{Z(\mathbf{u}) \le z_k | n(\mathbf{u})\} \le \text{Prob}\{Z(\mathbf{u}) \le z_{k'} | n(\mathbf{u})\}, \forall z_k \le z_{k'}$.

From these IK-derived conditional probabilities a class indicator can be simulated at each simulation node $\mathbf{u}$, the indicator of a category or of the class to which the continuous z-value belongs.

Note that at each node $(K - 1)$ indicator krigings are needed if K classes are considered, each kriging calling for its own indicator covariance model. In the case of a continuous variable $Z(\mathbf{u})$, the modeling task is considerably reduced if one adopts a median indicator model, whereby all $(K - 1)$ indicator covariance models are chosen proportional to the single covariance corresponding to the indicator defined by the median threshold value $z_k = M$ (Goovaerts, 1997, p.304; Chilès and Delfiner, 1999, p.384).

Recall that the individual indicator kriging results must be corrected to honor the constraints associated to expression (3.42) (Goovaerts, 1997, p.324). These order relation corrections are made before simulated values can be drawn from the IK-estimated conditional probabilities.

It can be shown that the indicator covariance models are reproduced, except for the impact of the order relation corrections.

Remarks

The indicator formalism was originally designed for categorical variables and later extended to continuous variables. As for simulation of several (say $K \ge 4$) categorical variables, *SISIM* should be used with care, as the number and magnitude of order relation corrections become prohibitive and the reproduction of the numerous indicator covariances becomes poor.

Hierarchy and spatial nesting of categories can be used to split the simulation of a large number K of categories into a series of independent simulations, each with a smaller number of categories (Maharaja, 2004). For example, the simulation of $K = 5$ lithofacies may be reduced to, first a simulation of the two dominant groups of facies ($K = 2$), followed by the simulation of the individual facies nested within each group, say $K = 3$ within any previously simulated first group and $K = 2$ for the second group.

3.9 Multiple-point simulation algorithms

The concept of multiple-point simulation was triggered by the failure of well established object-based algorithms to honor a large amount of local data. With object-based algorithms, also called Boolean algorithms, "objects" of given shape are dropped onto the simulation study area thus painting onto that area the desired shapes and patterns (Chilès and Delfiner, 1999, p.545; Stoyan *et al.*, 1987; Haldorsen and Damsleth, 1990; Lantuéjoul, 2002; Mallet, 2002). The object shape parameters, e.g. size, anisotropy, sinuosity, are made random thus making the simulation process stochastic. An iterative process is then applied for local data conditioning: objects are displaced, transformed, removed, replaced until a reasonable match is achieved. Object-based algorithms are ideal for building a training image with the required spatial structures and patterns, but they are notoriously difficult to condition to local data, particularly when these data are of small support volume, numerous and of diverse types. Conversely, pixel based algorithms are easy to condition because the simulation proceeds one pixel (point) at a time: modifying a single point-support value to match local data does not affect a whole object area around that point. But traditional pixel-based algorithms based on 2-point statistics could only reproduce a variogram or covariance model, failing to reproduce definite shapes and patterns.

3.9.1 Single normal equation simulation (SNESIM)

Without losing the data conditioning flexibility of a pixel-based procedure, one had to find a way around the variogram limitation. That variogram comes in only when building from kriging the local conditional probability distributions (see previous Section 3.8), hence the idea of collecting directly those distributions from training images that display the required spatial patterns. By so doing one would sidestep any variogram/covariance modeling and also any kriging. The probability distributions are chosen from the training image such as to match, exactly or approximately, the local conditioning data. More precisely, the training image is scanned to retrieve replicates of the conditioning data event; these replicates define a sub training population conditioned to the data from which the previous conditional distributions could be retrieved (Guardiano and Srivastava, 1993; Strebelle, 2002). The *SNESIM* algorithm reads conditional distributions from training images that could have been built using ideally suited non-conditional object-based algorithms, and progresses sequentially one pixel at a time thus capitalizing on the data conditioning ease of sequential simulation.

The main requisite, and a difficult one of the *SNESIM* algorithm, is a "rich" training image where enough exact replicates can be found for any conditioning data event encountered during the sequential simulation process. At any location, if not enough such replicates are found, some of the local conditioning data are dropped, allowing the possibility to find more replicates but at the cost of poorer data conditioning. That limitation becomes prohibitive if the simulation addresses too many categories ($K > 4$), or if the variable simulated is continuous. One must then revert to the *FILTERSIM* algorithm (Journel and Zhang, 2006; Zhang *et al.*, 2006) which accepts approximate replicates of the conditioning data event (described in Section 3.9.2).

The multiple-point (mp) sequential simulation algorithm whereby all conditional probabilities are read as corresponding proportions from a training image is called "Single Normal Equation SIMulation." This name recalls that any such proportion is in fact the result of a single indicator kriging (normal) equation, see relation (3.38).

The original *SNESIM* code (Guardiano and Srivastava, 1993) had to re-scan the training image anew at each simulation node to collect replicates of that node conditioning data event; it gave good results but was CPU-prohibitive. The breakthrough came with the introduction of the search tree concept which allowed for a single scan of the training image and smart storage in central memory of *all* resulting training proportions (Strebelle, 2002). These proportions are then directly read from the search tree during the course of sequential simulation. The roadblock of the *SNESIM* algorithm is not anymore one of CPU, it is the demand for a large and "rich" enough training image carrying enough replicates of most of the conditioning data events found in the course of simulation.

Details about the search tree-based *SNESIM* algorithm can be found in Strebelle (2000). The implementation of the *SNESIM* algorithm can be found in Section 8.2.1.

3.9.2 *Filter-based algorithm* (FILTERSIM)

A middle alternative between pixel-based and object-based algorithms is to cut into small pieces the objects or, better, a whole training image, then use those pieces in building a simulation making sure the pieces fit the conditioning data (Arpat, 2004;

Zhang, 2006; Journel and Zhang, 2006). The best analogy is perhaps that of building a puzzle, where each new piece patched onto the image being simulated must fit close-by previously placed pieces and original data. The search for pieces that would fit is speeded up by looking into bins containing previously classified "similar" looking pieces; say, one specific bin would contain all Ti pieces with some elements of sky in it, another bin would contain parts of trees and houses in it. As opposed to the puzzle game, any piece taken out of a bin is immediately replaced by an identical one thus no bin ever gets exhausted. Also the fit required is only approximate and it can be reconsidered later in the sequential simulation path.

Instead of putting down on the simulation field an entire Ti piece, only the central part of that piece can be patched down. That central part or patch size can be as small as the single central pixel value.

The critical key to the success of the *FILTERSIM* algorithm is the classification of local patterns of the training image into a not too large number of bins of "similar" looking patterns. That classification requires reducing any pattern to a small number of characteristic scores, say, sky with clouds or sky without clouds. In *FILTERSIM* these scores are defined through linear filters applied to the set of pixel values constituting the pattern (Schneiderman and Kanade, 2004). Next, one must define a distance between a conditioning data event and any such previous bin. This is needed to select the bin closest, that is with training patterns most similar, to the conditioning data event. Future research will undoubtedly suggest better pairs of (filters + distance) than that coded in this early version of the *FILTERSIM* code.

The *FILTERSIM* algorithm, originally designed for simulation of continuous variables, has been extended to categorical variables. However, and because the notion of linear filters does not extend naturally to categorical variables, we recommend using the categorical *FILTERSIM* approach only when it is absolutely necessary to simulate jointly a large number of categorical variables ($K > 4$). For a reasonable, hence small, number of categories the *SNESIM* approach is a better choice provided a corresponding large and varied (rich) training image is available.

Hierarchical simulation

Because of the difficulty in obtaining very large and rich training images, particularly in 3D, and because of the RAM demand of the corresponding search trees, it may not be feasible to apply the *SNESIM* algorithm to the joint simulation of more than $K = 4$ categories. That limitation is generally not a problem in earth sciences applications, since facies or rock types are often nested in each other, which allows decomposing the problem, see Maharaja (2004) and Walker (1984).

Consider, for example, the simulation of 7 facies, with facies #5 and #6 nested within facies #3, and facies #7 nested within facies #4. A first run of *SNESIM* with the modified four-facies training image depicting the spatial distribution of the four facies groups $A = 1$, $B = 2$, $C = 3 + 5 + 6$, $D = 4 + 7$ would yield simulated realizations of these four groups. Consider any one such realization and isolate the corresponding zones simulated as group C and group D; within the zones defined by group C use *SNESIM* with the proper Ti to simulate the spatial distribution of facies #3, #5 and #6; within zone D use *SNESIM* with yet another Ti to simulate the distribution of facies #4 and #7.

3.10 The nu/tau expression for combining conditional probabilities

Stochastic prediction is all about proposing a model for the probability distribution of possible outcomes of an unknown given all the data available. From such distribution model, one can simulate a set of outcomes for the unknown(s). The fundamental task is thus to determine the previous conditional distributions, a task particularly difficult when data of different types are present; data that are often redundant one with each and all others, and data whose information content goes much beyond a linear correlation with the unknown being assessed. Recent developments have uncovered a dormant general formulation of that general problem, one that lends itself remarkably to the modern mp approach to data integration (Bordley, 1982; Benediktsson and Swain, 1992; Journel, 2002; Polyakova and Journel, in press).

At this level of absolute generality, some notations are necessary; we will endeavor, however, to back these notations with intuitive examples.

Adopt the notation A for the unsampled RV, and the notations $D_i = d_i$, $i = 1, \ldots, n$ for the n data events, with capital letters denoting the RVs and the corresponding small case letters denoting any observed data value. In mp applications, the D_is are actually vectors involving multiple data locations, but we will keep the scalar notation D_i for simplicity.

The ultimate goal of probabilistic prediction is to evaluate the fully conditional probability:

$$\text{Prob}\{A = a | D_i = d_i, i = 1, \ldots, n\}, \qquad (3.43)$$

a function of the $(n + 1)$ values $(a; d_i, i = 1, \ldots, n)$.

If each data event D_i relates to a single location in space, say $d_i = z(\mathbf{u}_i)$, then a traditional two-point statistics such as the covariance suffices to relate any datum D_i to any other D_j or to the unknown A.

If each data event $\mathbf{D}_i$ involves jointly multiple data locations (it is then a vector) all related to the *same* attribute z which is also the A-attribute, then one could hope to find or build a Z-training image depicting the joint distribution of A and any vector of Z-values. Using such training image, the mp algorithms *SNESIM* and *FILTERSIM* could be implemented, see Sections 3.9.1 and 3.9.2.

However, in the general situation where each data event $\mathbf{D}_i$, in addition to being multiple-point, is also related to a different attribute, the task becomes insuperable. For example, $\mathbf{D}_1$ could be a mp pattern of facies indicator data as interpreted from well logs, $\mathbf{D}_2$ could be a set of seismic impedance data involving many locations in space but locations different from those related to $\mathbf{D}_1$; as for A it may relate to a third attribute, say porosity, at yet a different location (or set of locations).

The solution is again to "divide and conquer," decomposing the global data event $\mathbf{D} = \{D_i = d_i, i = 1, \ldots, n\}$ into n component data events D_i for which each of the individual conditional probabilities $\text{Prob}\{A = a | D_i = d_i\}$, $i = 1, \ldots, n$ could be evaluated by traditional two-point or by mp geostatistics or by any other means. The general problem is then that of recombining n individual probabilities into an estimate of the fully conditioned probability (Eq. (3.43)); this calls for determination of the integration function φ below:

$$\text{Prob}\{A = a | D_i = d_i, i = 1, \ldots, n\} = \varphi \left(\text{Prob}\{A = a | D_i = d_i\}, i = 1, \ldots, n \right).$$
$$(3.44)$$

Returning to the previous example:

$P(A|D_1)$ could be evaluated from a training image depicting the joint distribution of porosity (the A-attribute) and facies indicators (the D_1 categorical attribute),

$P(A|D_2)$ could be evaluated independently from calibration of porosity to a set of neighboring seismic impedance data (the D_2 categorical attribute),

it remains to combine these two partially conditioned probabilities accounting for the redundancy of seismic and facies data when it comes to evaluate porosity (A).

Fortunately there exists an *exact* decomposition formula of type (3.44), the so-called nu or tau expression. This expression has been known for some time (Bordley, 1982; Benediktsson and Swain, 1992), but its generality or exactitude had not been established until recently, nor was its importance for data integration fully recognized.

Warning

All probabilities in expressions (3.43) and (3.44) are functions of the $(n + 1)$ values a and d_i, more if $\mathbf{d}_i$ is a mp vector of data values. However, for simplicity we will use the short notations $P\{A|D\}$ and $P\{A|D_i\}$ whenever there is no risk of confusion.

Why probabilities?

Before developing the expression of the compositing function φ, one should answer the question whether a probabilistic approach is the most appropriate for this data integration problem. The answer lies in the notation (3.44) itself:

- probabilities provide a unit-free, standardized $[0, 1]$, coding of information, across all data types, which facilitates the task of data integration;
- as opposed to a deterministic estimate of A, each elementary probability $P(A = a|D_i = d_i)$ includes both the d_i information content and the uncertainty of its contribution to evaluating $A = a$.

The nu/tau expression

Consider the probability-into-distance transform of each individual probability:

$$x_0 = \frac{1 - P(A)}{P(A)}, \quad x_1 = \frac{1 - P(A|D_1)}{P(A|D_1)}, \quad x_n = \frac{1 - P(A|D_n)}{P(A|D_n)}, \tag{3.45}$$

all valued in $[0, +\infty]$.

$P(A) = P(A = a)$ is the prior probability of event $A = a$ occurring, "prior" to knowing any of the n data $D_i = d_i$, x_0 is the prior distance to $A = a$ occurring, equal to zero if $P(A) = 1$, equal to ∞ if $P(A) = 0$, and similarly for each of the elementary distance x_i.

We will use the notation: $1 - P(A|D_i) = P(\widetilde{A}|D_i)$, where $\widetilde{A}$ stands for non$A$.

The distance x to $A = a$ occurring given *jointly* all n data is given by the nu, expression:

$$\frac{x}{x_0} = \prod_{i=1}^{n} \nu_i \frac{x_i}{x_0} = \nu_0 \cdot \prod_{i=1}^{n} \frac{x_i}{x_0}, \quad \text{with: } \nu_i \geq 0$$

or equivalently the tau expression:

$$\frac{x}{x_0} = \prod_{i=1}^{n} (\frac{x_i}{x_0})^{\tau_i}, \quad \text{with: } \tau_i \in [-\infty, +\infty] \tag{3.46}$$

with

$$\nu_i = \left(\frac{x_i}{x_0}\right)^{\tau_i - 1}, \quad \text{or: } \tau_i = 1 + \frac{\log \nu_i}{\log \frac{x_i}{x_0}}$$

and

$$\nu_0 = \prod_{i=1}^{n} \nu_i \in [0, +\infty]. \tag{3.47}$$

Recall that:

$$x = \frac{P\left(\widetilde{A}|D_i, \ldots, D_n\right)}{P\left(A|D_1, \ldots, D_n\right)}$$

thus

$$P\left(A|D_1, \ldots, D_n\right) = \frac{1}{1+x} \in [0, 1]. \tag{3.48}$$

The nu/tau expressions give the fully conditioned relative distance x/x_0 as function of the n elementary relative distances x_i/x_0. Recall that these n elementary distances are assumed known, the only problem addressed by the two equivalent relations (3.46) is that of combining the elementary distances into the fully conditioned distance x. Expression (3.46) shows the combination function to be a weighted product as opposed to an approach by indicator kriging that considers weighted linear combination; see Section 3.6.5. The relative distances carry the information content of each elementary data event D_i; the tau or nu weights account for the additional information (beyond redundancy) carried by the various data events as to evaluating the probability for $A = a$.

The exact expression of the ν-parameters is (Polyakova and Journel, in press):

$$\nu_i = \frac{\frac{P(D_i|\widetilde{A}, \overline{D_{i-1}})}{P(D_i|A, \overline{D_{i-1}})}}{\frac{P(D_i|\widetilde{A})}{P(D_i|A)}} \in [0, +\infty], \quad \nu_1 = 1. \tag{3.49}$$

Similarly for the tau parameters (Krishnan, 2004):

$$\tau_i = \frac{\log \frac{P(D_i|\widetilde{A}, \overline{D_{i-1}})}{P(D_i|A, \overline{D_{i-1}})}}{\log \frac{P(D_i|\widetilde{A})}{P(D_i|A)}} \in [-\infty, +\infty], \quad \tau_1 = 1, \tag{3.50}$$

where $\overline{D_{i-1}} = \{D_j = d_j, j = 1, \ldots, i - 1\}$ denotes the set of all data events considered before the ith data event $D_i = d_i$.

Prob$(D_i|A)$ is the (likelihood) probability of observing the datum value $D_i = d_i$ given the outcome $A = a$, Prob$(D_i|\widetilde{A})$ is the probability of observing the same data but given $\widetilde{A}$, thus the ratio $\frac{\text{Prob}(D_i|\widetilde{A})}{\text{Prob}(D_i|A)}$ appearing in the denominator of the ν_i or τ_i expression can be read as a measure of how datum $D_i = d_i$ discriminates A from $\widetilde{A}$. The ratio appearing in the numerator is the same discrimination measure but in presence of all previous data considered $\overline{D_{i-1}} = \{D_1 = d_1, \ldots, D_{i-1} = d_{i-1}\}$. The unit value $\nu_i = \tau_i = 1$ would correspond to full information redundancy between the data event $D_i = d_i$ and the previously considered data $\overline{D_{i-1}}$. Thus the parameter values $|1 - \nu_i|$ or $|1 - \tau_i|$ could be read as the additional information content brought by D_i above the previous data $\overline{D_{i-1}}$ as to discriminating $A = a$ from $A = $ non-a.

Note that the single correction parameter ν_0 is data sequence-independent; also the case $\nu_0 = 1$ is more general than $\nu_i = 1, \forall i$; it encompasses the complex case of data redundancies ($\nu_i \neq 1$) that cancel each other globally into $\nu_0 = 1$.

Tau or nu model?

Recall that all expressions above (Eq. (3.43) to Eq. (3.50)) are data values-dependent notwithstanding their short notation, say, $P(A|D_i)$ should be read $P(A = a|D_i = d_i)$; similarly the elementary distance x_i is both a and d_i values-dependent.

If the ν_i, τ_i parameters are actually evaluated, e.g. from training image, and made data values-dependent, the two expressions in Eq. (3.47) are equivalent. One would then prefer the nu-formulation because it puts forward a single correction parameter $\nu_0(a, d_i; i = 1, \ldots, n)$ which is independent of the data sequence $D_1, D_2, \ldots, D_n$. Also, evaluation of the τ_i parameter associated to a non-informative datum such that $P(D_i|\widetilde{A}) \approx P(D_i|A)$ would run into problems because of a division by a log ratio close to zero, see Eq. (3.50).

However, if the ν_i, τ_i parameters are assumed constant, independent of the $(a, d_i; i = 1, \ldots, n)$ values, then the tau formulation should be preferred. Indeed, consider the case of only two data events with the two different sets of data values:

$$\{D_1 = d_1, D_2 = d_2\} \text{ and } \left\{D_1 = d_1', D_2 = d_2'\right\}.$$

- The nu model with constant (homoscedastic) ν_0 parameter value is written:

$$\frac{x}{x_0} = \nu_0 \cdot \frac{x_1}{x_0} \cdot \frac{x_2}{x_0} \text{ for data set } \{d_1, d_2\}$$

$$\frac{x'}{x_0} = \nu_0 \cdot \frac{x_1'}{x_0} \cdot \frac{x_2'}{x_0} \text{ for data set } \left\{d_1', d_2'\right\}$$

where x, x_1, x_2 are the distances corresponding to $\{d_1, d_2\}$ and x', x_1', x_2' are the distances corresponding to $\left\{d_1', d_2'\right\}$. Conditional distances are data values-dependent, as opposed to the prior distance $x_0 = x_0'$. Therefore,

$$\frac{x'}{x} = \frac{x_1'}{x_1} \cdot \frac{x_2'}{x_2}, \quad \forall \nu_0.$$

The parameter ν_0 is seen to be ineffective.

- Conversely, the tau model with constant τ_1, τ_2 parameter values is written:

$$\log \frac{x}{x_0} = \tau_1 \cdot \log \frac{x_1}{x_0} + \tau_2 \cdot \log \frac{x_2}{x_0} \text{ for data set } \{d_1, d_2\}$$

$$\log \frac{x'}{x_0} = \tau_1 \cdot log \frac{x_1'}{x_0} + \tau_2 \cdot \log \frac{x_2'}{x_0} \text{ for data set } \left\{d_1', d_2'\right\}.$$

Thus:

$$\log \frac{x'}{x} = \tau_1 \cdot \log \frac{x_1'}{x_1} + \tau_2 \cdot \log \frac{x_2'}{x_2}, \quad \text{or equivalently}$$

$$\frac{x'}{x} = \frac{x_1'}{x_1}^{\tau_1} \cdot \frac{x_2'}{x_2}^{\tau_2}.$$

The tau parameters, although data values-independent, remain effective unless $\tau_1 = \tau_2 = \nu_0 = 1$.

This latter property of the tau expression, remaining effective even if the τ_is are considered data values-independent, make the tau expression (3.46) a convenient heuristic to weight more certain data events. It suffices to make $\tau_i > \tau_j > 0$ to give more importance to data event D_i as compared to data event D_j, whatever the actual data values (d_i, d_j). That heuristic utilization of the tau model completely misses the main contribution of the nu/tau expression which is the quantification of data redundancy for any specific set of values $(a, d_i; i = 1, \ldots, n)$.

SGeMS proposes a utility program, *NU-TAU MODEL* see Section 9.5, to combine prior probabilities using either the nu or the tau expression (3.46) with as input data values-dependent nu or tau parameters. However, in programs *SNESIM* and *FILTERSIM*, only the tau expression is allowed with tau parameters input as data values-independent constant values.

3.11 Inverse problem

A major topic not directly addressed by the SGeMS software is that of integration of difficult data D expressed as a non-analytical, non-linear function ψ of a large number of values $z(\mathbf{u}_\alpha)$ being simulated :

$$D = \psi(z(\mathbf{u}_\alpha), \alpha = 1, \ldots, n)$$

The simulated fields $\{z^{(l)}(\mathbf{u}), \mathbf{u} \in S\}$, $l = 1, \ldots, L$, must be such that they all reproduce such data, i.e.

$$D^{(l)} = \psi(\{z^{(l)}(\mathbf{u}), \alpha = 1, \ldots, n) \approx D \quad \forall l = 1, \ldots, L$$

where the function ψ is known, although typically only through an algorithm such as a flow simulator.

SGeMS provides realizations $\{z^{(l)}(\mathbf{u}), \mathbf{u} \in S\}$ that can be selected, combined, perturbed and checked to fit approximately the data D. This is known as the general "inverse problem" (Tarantola, 2005); see Hu *et al.* (2001) and Caers and Hoffman (2006) for a geostatistical perspective.

4

Data sets and SGeMS EDA tools

This chapter presents the data sets used to demonstrate the geostatistics algorithms in the following chapters. It also provides an introduction to the exploratory data analysis (EDA) tools of the SGeMS software.

Section 4.1 presents the two data sets: one in 2D and one in 3D. The smaller 2D data set is enough to illustrate the running of most geostatistics algorithms (kriging and variogram-based simulation). The 3D data set, which mimics a large deltaic channel reservoir, is used to demonstrate the practice of these algorithms on large 3D applications; this 3D data set is also used for EDA illustrations.

Section 4.2 introduces the basic EDA tools, such as histogram, Q-Q (quantile–quantile) plot, P-P (probability–probability) plot and scatter plot.

4.1 The data sets

4.1.1 The 2D data set

This 2D data set is derived from the published Ely data set (Journel and Kyriakidis, 2004) by taking the logarithm of all positive values and discarding the negative ones. The original data are elevation values in the Ely area, Nevada. The corresponding SGeMS project is located at `DataSets/Ely1.prj`. This project contains two SGeMS objects: `Ely1_pset` and `Ely1_pset_samples`.

- The `Ely1_pset` object is a point set grid with 10 000 points, constituting a reference (exhaustive) data set. This point set grid holds three properties: a local varying mean data ("lvm"), the values of the primary variable ("Primary") and the values of a co-located secondary property ("Secondary"). The point set grid and its properties are given in Fig. 4.1a–d. This object can be used to hold properties obtained from kriging algorithms or stochastic simulations.

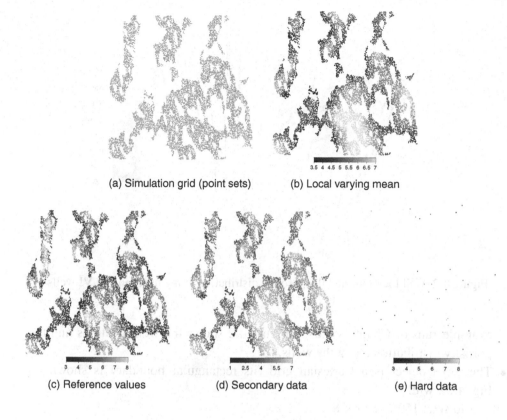

(a) Simulation grid (point sets) (b) Local varying mean

(c) Reference values (d) Secondary data (e) Hard data

Figure 4.1 The Ely data set

- The `Ely1_pset_samples` object provides 50 well data ("samples"), which can be used as hard primary data to constrain the geostatistical estimations or simulations. These data, shown in Fig. 4.1e, were sampled from the reference data set (Fig. 4.1c).

4.1.2 The 3D data set

The 3D data set retained in this book is extracted from a layer of Stanford VI, a synthetic data set representing a fluvial channel reservoir (Castro, 2007). The corresponding SGeMS project is located at `DataSets/stanford6.prj`. This project contains three SGeMS objects: `well`, `grid` and `container`.

- The `well` object contains the well data set. There is a total of 26 wells (21 vertical wells, four deviated wells and one horizontal well). The six properties associated with these wells are bulk `density`, a binary `facies` indicator (sand channel or mud floodplain), `P-wave impedance`, `P-wave velocity`, `permeability` and `porosity`. These data will be used as hard or soft conditioning data in the

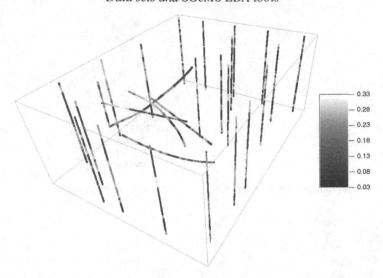

Figure 4.2 Well locations and the porosity distribution along the Stanford VI wells

example runs of Chapters 7 to 9. Figure 4.2 shows the well locations and the porosity distribution along the wells.

- The `grid` object is a Cartesian grid (its rectangular boundary is shown on Fig. 4.2), with
  - grid size: 150 × 200 × 80,
  - origin point at (0,0,0),
  - unit cell size in each x/y/z direction.

This reservoir grid holds the following two variables.

1. Probability data. The facies probability data were calibrated from the original seismic impedance data using the well data (*facies* and *P-wave impedance*). Two sand probability cubes (properties `P(sand|seis)` and `P(sand|seis)_2`) are provided: the first displays sharp channel boundaries (best quality data, see Fig. 4.3a); the second displays more fuzzy channel boundaries (poor quality data, see Fig. 4.3b). These probability data will be used as soft data to constrain the facies modeling.

2. Region code. Typically a large reservoir would be divided into different regions with each individual region having its own characteristics, for instance, different channel orientations and channel thickness. The regions associated with the Stanford VI reservoir are rotation regions (property `angle`) corresponding to different channel orientations (Fig. 4.4), and affinity (scaling) regions (property `affinity`) corresponding to different channel thicknesses (Fig. 4.5). Each rotation region is labeled with an indicator number, and is assigned a rotation angle value, see Table 4.1. The affinity indicators and the attached affinity values are given in Table 4.2. An affinity

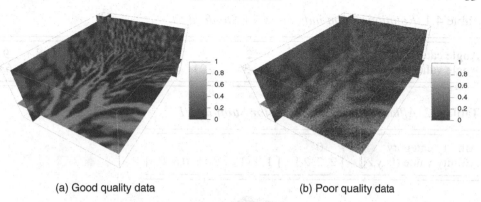

(a) Good quality data   (b) Poor quality data

Figure 4.3 Two Stanford VI sand probability cubes

Figure 4.4 Angle indicator cube

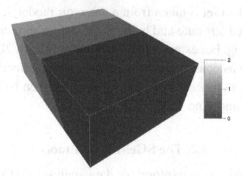

Figure 4.5 Affinity indicator cube

value must be assigned to each x/y/z direction; the larger the affinity value, the thicker the channel in that direction.

- The `container` object is composed of all the reservoir nodes located inside the channels, hence it is a point-set with (x,y,z) coordinates. The user can perform geostatistics on this channel container, for example, to estimate the within-channel petrophysical properties. In Fig. 4.6 the channel container is represented by all nodes with value 1 (gray), and the non-reservoir area is in black.

Table 4.1 *Rotation region indicators for Stanford VI*

Angle category	0	1	2	3	4	5	6	7	8	9
Angle value (degree)	−63	−49	−35	−21	−7	7	21	35	49	63

Table 4.2 *Affinity region indicators for Stanford VI*

Affinity category	0	1	2
Affinity value ([x,y,z])	[2, 2, 2]	[1, 1, 1]	[0.5, 0.5, 0.5]

Figure 4.6 Stanford VI channel container (gray nodes)

Although this 3D data set is taken from a reservoir model, it could represent any 3D spatially distributed attribute and be used for testing applications in other fields than reservoir modeling. For example, one can interpret each 2D horizontal layer of the seismic data cube as coarse satellite measurements defined over the same area but recorded at different times. The application would then be modeling landscape change in both space and time.

4.2 The SGeMS EDA tools

SGeMS provides some useful exploratory data analysis (EDA) tools, such as histogram, quantile–quantile (Q-Q) plot, probability–probability (P-P) plot, scatter plot, variogram and cross-variogram calculation and modeling. In this chapter, the first four elementary tools are presented; the (cross-)variogram calculation and modeling tool is described in the next chapter.

All the EDA tools can be invoked through the *Data Analysis* menu from the main SGeMS graphical interface. Once a specific tool is selected, the corresponding SGeMS window is popped up. The EDA tool window is independent of the main SGeMS interface, and the user can have multiple windows for each EDA tool.

4.2.1 Common parameters

The interface for all EDA tools presented in this chapter has three panels; see also Figs 4.7 to 4.9.

1. **Parameter Panel**   The user selects in this panel the properties to be analyzed and the display options. This panel has two pages: "Data" and "Display Options", the latter being common to all EDA tools.
2. **Visualization Panel**   This panel shows the graphic result of the selected statistics.
3. **Statistics Panel**   This panel displays some relevant summary statistics.

In the lower part of the main interface, there are two buttons: *Save as Image* and *Close*. The *Save as Image* button is used to save a graphical result (for example a histogram) into a picture data file in either "png", "bmp" or "ps" (Postscript) format. The *Close* button is used to close the current interface.

Parameters description

The parameters of the "Display Options" page are described below.

- **X Axis**   Controls for the X axis for variable 1. Only the property values between "Min" and "Max" are displayed in the plot; values less than "Min" or greater than "Max" still contribute to the statistical summaries. The default values of "Min" and "Max" are the minimum and maximum of the selected Property. The **X Axis** can be set to a logarithmic scale by marking the corresponding check box. This option is valid only when all the property values are larger than zero.
- **Y Axis**   Controls for the Y axis for variable 2. The previous remarks apply.

The user can modify the parameters through either the keyboard or the mouse. Any modification through the mouse will instantly reflect on the visualization or the summary statistics.

Warning: the change through the keyboard must be activated by pressing the "Enter" key.

4.2.2 Histogram

The *histogram* tool creates a visual output of the frequency distribution, and displays some summary statistics, such as the mean and variance of the selected variable. The *histogram* tool is activated by clicking *Data Analysis → Histogram*. Although the program will automatically scale the histogram, the user can set the histogram limits in the *Parameter Panel*. The main *histogram* interface is given in Fig. 4.7, and the parameters of the *Data* page are listed below.

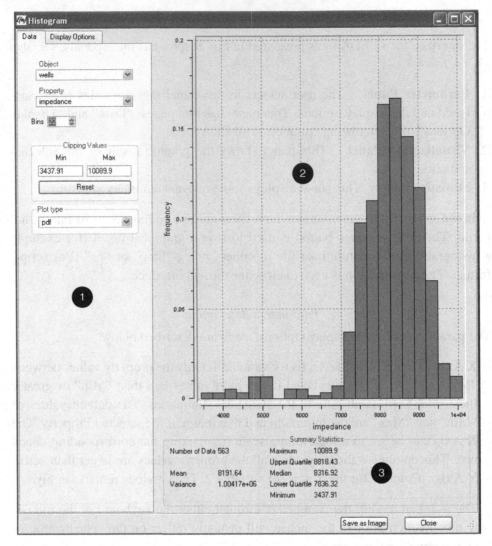

Figure 4.7   Histogram interface [1]: Parameter Panel; [2]: Visualization Panel; [3]: Statistics Panel

Parameters description

- **Object**   A Cartesian grid or a point-set containing the variables under study.
- **Property**   The variable to study.
- **Bins**   The number of classes. The user can change this number through the keyboard, or by clicking the scroll bar. Any value change will be instantly reflected on the histogram display.
- **Clipping Values**   Statistical calculation settings. All values less than "Min" and greater than "Max" are ignored, and any change of "Min" and "Max" will

affect the statistics calculation. The default values of "Min" and "Max" are the minimum and maximum of the selected **Property**. After modifying "Min" and/or "Max", the user can go back to the default setting by clicking "Reset".

- **Plot type** The user can choose to plot a frequency histogram ("pdf"), a cumulative histogram ("cdf") or both.

4.2.3 Q-Q plot and P-P plot

The Q-Q plot compares equal p-quantile values of two distributions; the P-P plot compares the cumulative probability distributions of two variables for equal threshold values. The two variables need not be in the same object or have the same number of data. The Q-Q plot and P-P plot are combined into one program, which can be invoked from *Data Analysis* → *QQ-plot*. This EDA tool generates both a graph in the *Visualization Panel* and some summary statistics (mean and variance for each variable) in the *Statistics Panel*, see Fig. 4.8. The parameters in the "Data" page are listed below.

Parameters description

- **Analysis Type** Algorithm selection. The user can choose either a Q-Q plot or a P-P plot.
- **Variable 1** The variable selection for the X axis. The user must choose first an object, then the property name.
- **Clipping Values for Variable 1** All values strictly less than "Min" and strictly greater than "Max" are ignored; any change of "Min" and "Max" will affect the statistics calculation. The user can go back to the default setting by clicking "Reset".
- **Variable 2** The variable selection for the Y axis. The user must choose first an **Object**, then the **Property** name. Note that **Variable 2** and **Variable 1** might be from different objects.
- **Clipping Values for Variable 2** Remarks similar to those for **Clipping Values for Variable 1**.

4.2.4 Scatter plot

The *scatterplot* tool (executed by clicking *Data Analysis* → *Scatter-plot*) is used to compare two variables by displaying their bivariate scatter plot and some statistics. All available data pairs are used to compute the summary statistics, such as the correlation coefficient, the mean and variance of each variable (see part [C] in Fig. 4.9). To avoid a crowded figure in the Visualization Panel, only up to 10 000 data pairs are displayed in the scatter plot. The parameters in the "Data" page are listed below.

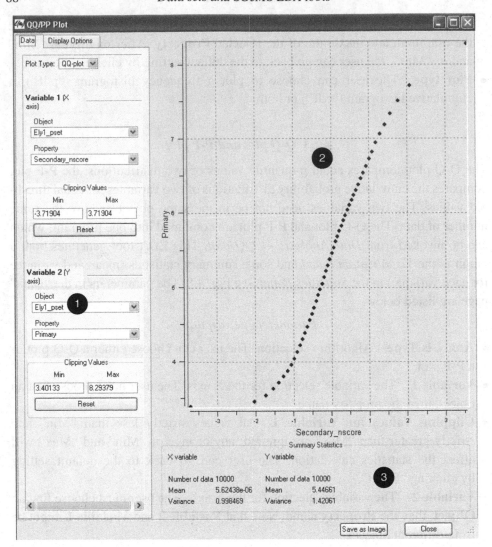

Figure 4.8 Q-Q plot interface [1]: Parameter Panel; [2]: Visualization Panel; [3]: Statistics Panel

Parameters description

- **Object** A Cartesian grid or a point-set containing the variables under study. This **Object** must contain at least two properties.
- **Variable 1** The variable property listed in the **Object** above. This variable is associated with the X axis.
- **Clipping Values for Variable 1** All values strictly less than "Min" and strictly greater than "Max" are ignored, and any change of "Min" and "Max" will affect the statistics calculation. The user can go back to the default setting by clicking "Reset". If **Variable 1** has more than 10 000 data, then the "Reset" button can be

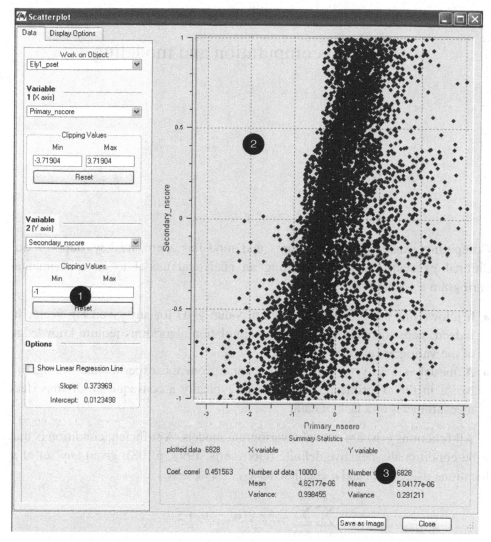

Figure 4.9 Scatter plot interface [1]: Parameter Panel; [2]: Visualization Panel; [3]: Statistics Panel

used to generate a new scatter plot with a re-sampled set of data pairs containing up to 10 000 data.

- **Variable 2**   The variable property listed in the upper **Object**. This variable is associated with the Y axis.
- **Clipping Values for Variable 2**   Remarks similar to those for **Variable 1**.
- **Options**   The choice of visualizing the least square line fit in the scatter plot. The slope and the intercept are given below check box "Show Least Square Fit". This option is valid only when the two variables are displayed with the arithmetical scale.

5

Variogram computation and modeling

Computing experimental variograms and modeling them are key steps of traditional geostatistical studies. Fitting an analytical model to an experimental variogram achieves two purposes.

- It allows one to compute a variogram value $\gamma(\mathbf{h})$ for any given lag vector $\mathbf{h}$. Indeed, geostatistical estimation and simulation algorithms require knowledge of the variogram at arbitrary lags.
- A model is a way to filter out the noise from the experimental variogram. Noise in the experimental variogram is typically a consequence of imperfect measurements or a lack of data.

All functions $g(\mathbf{h})$ are not valid variogram models. A sufficient condition is that g be conditionally negative definite (Goovaerts, 1997, p.108): given any set of n locations $\mathbf{u}_1, \ldots, \mathbf{u}_n$, and n coefficients $\lambda_1, \ldots, \lambda_n \in \mathbb{R}$,

$$\sum_{\alpha=1}^{n} \sum_{\beta=1}^{n} \lambda_\alpha \lambda_\beta \, g(\mathbf{u}_\alpha - \mathbf{u}_\beta) \leq 0$$

under the condition $\sum_{\alpha=1}^{n} \lambda_\alpha = 0$.

SGeMS supports four basic analytical variogram models, and any positive linear combination of these variograms. The four analytical (semi-)variograms are as follows, in their isotropic form.

Nugget effect model

$$\gamma(\mathbf{h}) = \begin{cases} 0 & \text{if } \|\mathbf{h}\| = 0 \\ 1 & \text{otherwise.} \end{cases} \tag{5.1}$$

A pure nugget effect model for a variable $Z(\mathbf{u})$ expresses a lack of (linear) dependence between variables $Z(\mathbf{u})$ and $Z(\mathbf{u} + \mathbf{h})$.

Spherical model with range a

$$\gamma(\mathbf{h}) = \begin{cases} \frac{3}{2}\frac{\|\mathbf{h}\|}{a} - \frac{1}{2}(\frac{\|\mathbf{h}\|}{a})^3 & \text{if } \|\mathbf{h}\| \leq a \\ 1 & \text{otherwise.} \end{cases} \tag{5.2}$$

Exponential model with *practical* **range** a

$$\gamma(\mathbf{h}) = 1 - \exp\left(\frac{-3\|\mathbf{h}\|}{a}\right). \tag{5.3}$$

Gaussian model with *practical* **range** a

$$\gamma(\mathbf{h}) = 1 - \exp\left(\frac{-3\|\mathbf{h}\|^2}{a^2}\right). \tag{5.4}$$

All these models are monotonously increasing and bounded by 1: $0 \leq \gamma(\mathbf{h}) \leq 1, \forall \mathbf{h}$. In the case of the exponential and Gaussian models, the upper bound (the *sill*) is reached asymptotically, and the distance $\|\mathbf{h}\|$ at which 95% of the sill is reached is called the *practical* range.

The covariance counterpart of the above four models is given by

$$C(\mathbf{h}) = C(\mathbf{0}) - \gamma(\mathbf{h}), \quad \text{with } C(\mathbf{0}) = 1.$$

In SGeMS a variogram model: $\gamma(\mathbf{h}) = c_0\gamma^{(0)}(\mathbf{h}) + \sum_{l=1}^{L} c_l\gamma^{(l)}(\mathbf{h})$ is characterized by the following parameters:

- a nugget effect $c_0\gamma^{(0)}$ with nugget constant $c_0 \geq 0$;
- the number L of *nested* structures. Each structure $c_l\gamma^{(l)}(\mathbf{h})$ is then defined by:
  - a variance contribution $c_l \geq 0$,
  - the type of the variogram: spherical, exponential or Gaussian,
  - an anisotropy, characterized by an ellipsoid with three directions and the ranges along each direction, see Section 2.5. Note that each nested structure can have a different anisotropy.

Example Consider the variogram model $\gamma(\mathbf{h}) = 0.3\gamma^{(0)}(\mathbf{h}) + 0.4\gamma^{(1)}(\mathbf{h}) + 0.3\gamma^{(2)}(\mathbf{h})$, with:

- $\gamma^{(0)}(\mathbf{h})$ a pure nugget effect with sill 0.3;
- $\gamma^{(1)}(\mathbf{h})$ an anisotropic spherical variogram with major range 40, medium range 20 and minor range 5, and angles $\alpha = 45°$ (azimuth), $\beta = 0$ (dip), $\theta = 0$ (rake);
- $\gamma^{(2)}(\mathbf{h})$ an isotropic exponential variogram of range 200.

SGeMS would save that model as the following XML file:

```
<Variogram nugget="0.3" structures_count="2" >
    <structure_1 contribution="0.4" type="Spherical" >
        <ranges max="40" medium="20" min="5" />
        <angles x="45" y="0" z="0" />
    </structure_1>
    <structure_2 contribution="0.3" type="Exponential" >
        <ranges max="200" medium="200" min="200" />
        <angles x="0" y="0" z="0" />
    </structure_2>
</Variogram>
```

5.1 Variogram computation in SGeMS

Although the text only refers to variograms, SGeMS can also compute covariances, correlograms and cross-variograms (see Goovaerts (1997) for definitions of these correlation measures).

To bring up the SGeMS variogram module, select the *Variogram* entry from the *Data Analysis* menu. Variogram computation is done in three steps.

1. Select the *head* and *tail* variables whose (cross-)variogram will be computed (Fig. 5.1). The computed variogram will measure the variability between the two variables $Z_{head}(\mathbf{u} + \mathbf{h})$ and $Z_{tail}(\mathbf{u})$. To compute the auto-variogram of a variable Z, select the same variable for both the head and the tail.
2. Input the necessary parameters, such as the directions in which the variogram should be computed and the number of lags to use (Fig. 5.2a and Fig. 5.2b). The parameters required will differ, depending on whether the head and tail variables are defined on a set of points (i.e. with no pre-defined spatial structure) or on a Cartesian grid.
3. Display the results (Fig. 5.3). At that point, it is also possible to model the computed experimental variograms (see Section 5.2).

After completing each step, click the *Next* button to go to the next step.

5.1.1 Selecting the head and tail properties

Figure 5.1 shows the interface to select the *head* and *tail* variables whose (cross-)variogram $\gamma\ (Z_{head}(\mathbf{u} + \mathbf{h}), Z_{tail}(\mathbf{u}))$ will be computed.

Both head and tail variables must belong to the same object (i.e. to the same set of points or the same Cartesian grid). Use menu item *Objects → Copy Property* to transfer a property between two objects.

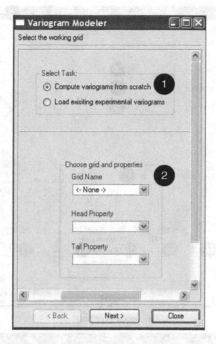

Figure 5.1 Interface to select the head and tail variables for the variogram computation

Description of the interface

1. **Select Task** Choose whether to compute a new variogram, or load an existing experimental variogram from a file.
2. **Choose grid and properties** Select the object that contains the head and tail properties and choose the head and tail properties. Select the same property for both head and tail to compute a univariate variogram, or two different properties to compute their cross-variogram.

5.1.2 Computation parameters

This step prompts for the number of lags at which to compute the experimental variogram, and the directions along which to compute that variogram. It is also possible to consider a type of correlation measure different from the variogram, for example a covariance or correlogram. The lags and directions along which to compute the variogram are input differently, depending on the type of the object (set of points or Cartesian grid) that holds the head and tail properties.

Parameters can be saved and loaded using the *Load Parameters* and *Save* buttons visible at the top of Fig. 5.2b.

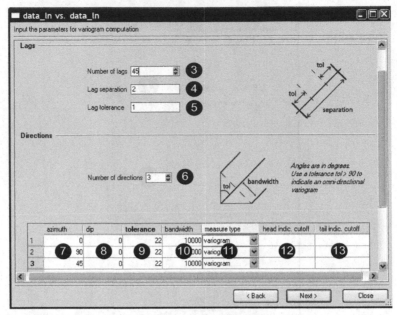

a Parameters for point-set

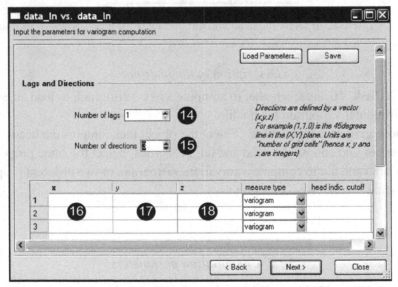

b Parameters for Cartesian grid

Figure 5.2 Parameters for the variogram computation

Parameters for data defined on a set of points

Given a number L of lags, a lag separation a and a set of K unit vectors $\mathbf{v}_1, \ldots, \mathbf{v}_K$, SGeMS will compute the following experimental variogram values.

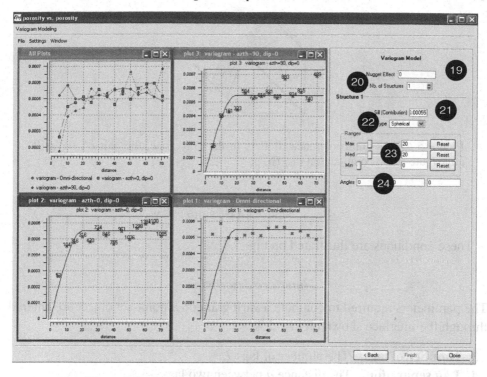

Figure 5.3 Variogram plots display and modeling interface

$$\begin{matrix} \gamma(a\mathbf{v}_1) & \ldots & \gamma(a \cdot L\mathbf{v}_1) \\ \vdots & \vdots & \vdots \\ \gamma(a\mathbf{v}_k) & \ldots & \gamma(a \cdot L\mathbf{v}_k) \end{matrix}$$

In a *point-set* object, data do not necessarily follow a regular spatial pattern. Because of that lack of structure, it is unlikely to find enough pairs of data separated by the same vector $\mathbf{h}$. Hence the need for a tolerance on the norm of $\mathbf{h}$ and its direction to compute a variogram on a point-set object.

The tolerance on $\mathbf{h}$ is characterized by three parameters:

- a lag tolerance ϵ,
- an angle $0 \leq \alpha_{\text{tol}} < 90°$,
- a bandwidth w,

such that two points A, B contribute to the computation of $\gamma(\mathbf{h})$ if:

$$\Big| \|\mathbf{AB}\| - \|\mathbf{h}\| \Big| \leq \epsilon$$

and, calling $\theta = (\mathbf{h}, \mathbf{AB})$ the angle between $\mathbf{h}$ and $\mathbf{AB}$,

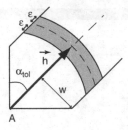

Figure 5.4 Variogram tolerance. If B is in the gray zone, the pair (A, B) will
contribute to the computation of $\gamma(\mathbf{h})$

$$\theta \leq \alpha_{\text{tol}} \quad \text{and} \quad \|\mathbf{AB}\| \sin(\theta) \leq w.$$

These conditions are illustrated on Fig. 5.4.

Interface description

The parameters required to compute a variogram on a point-set object are entered
through the interface shown on Fig. 5.2a.

3. **Number of lags**   The number of lags L.
4. **Lag separation**   The distance a between two lags.
5. **Lag tolerance**   The tolerance ϵ around the lag separation.
6. **Number of directions**   The number K of directions along which to compute
   the experimental variograms, each with the same number of lags. Each direction $\mathbf{v}_k$ ($k = 1, \ldots, K$) is characterized by two angles (items **7** and **8**) and a
   tolerance (items **9** and **10**).
7. **Azimuth**   The azimuth, in degrees, of vector $\mathbf{v}_k$ (Fig. 2.24).
8. **Dip**   The dip, in degrees, of vector $\mathbf{v}_k$ (Fig. 2.25).
9. **Tolerance**   The tolerance angle α_{tol}, in degrees. Specify an angle greater than
   90° to compute an omni-directional variogram.
10. **Bandwidth**   The bandwidth parameter w (Fig. 5.4).
11. **Measure type**   The measure of bivariate spatial correlation. The options are:
    variogram, indicator variogram, covariance and correlogram. If indicator var-
    iogram is selected, the head and tail values must be coded into indicators, see
    items **12** and **13** below.
12. **Head indicator cutoff**   This parameter is used only if the measure type
    (item **11**) is *indicator variogram*. It is the threshold z_t defining the indica-
    tor coding of the head variable z. If $z \leq z_t$, the indicator value is 1; it is 0
    otherwise.
13. **Tail indicator cutoff**   This parameter is used only if the measure type
    (item **11**) is *indicator variogram*. It is the threshold z_t defining the indicator
    coding of the tail variable z. If $z \leq z_t$, the indicator value is 1; it is 0 otherwise.

If computing an indicator auto-variogram, i.e. if the head and tail variables are the same, the head and tail indicator cutoffs must be the same.

Note Categorical indicator variogram should be calculated directly from input indicator data.

Parameters for data defined on a Cartesian grid

Given a number L of lags, a lag separation a and a set of K vectors $\mathbf{v}_1, \ldots, \mathbf{v}_K$, (these vectors can have a norm different from 1) SGeMS will compute the following experimental variogram values.

$$
\begin{matrix}
\gamma(a\mathbf{v}_1) & \ldots & \gamma(a \cdot L\mathbf{v}_1) \\
\vdots & \vdots & \vdots \\
\gamma(a\mathbf{v}_k) & \ldots & \gamma(a \cdot L\mathbf{v}_k)
\end{matrix}
$$

Contrary to the point-set case, there is no need to specify a distance and direction tolerance: since all data locations follow a regular pattern, multiple pairs are guaranteed to be available, unless a lot of data are missing.

Interface description

The parameters required to compute a variogram on a Cartesian grid are entered through the interface shown on Fig. 5.2b.

14. **Number of lags** The number of lags L.
15. **Number of directions** The number K of directions along which to compute the experimental variograms, each with the same number of lags. Each vector $\mathbf{v}_k$ is specified by its integer coordinates in the grid coordinate system, see items **16**, **17** and **18**, and Fig. 5.5.
16. **x** The X coordinate of vector $\mathbf{v}_k$. It is expressed in number of grid cells. So if the grid cells are 10 m long in the X direction, $x = 3$ means 30 m in the X direction.
17. **y** The Y coordinate of vector $\mathbf{v}_k$. It is expressed in number of grid cells, see item **16**.
18. **z** The Z coordinate of vector $\mathbf{v}_k$. It is expressed in number of grid cells, see item **16**.

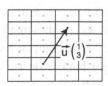

Figure 5.5 A vector in grid cell coordinates. The coordinates of $\mathbf{v}$ are $x = 1$, $y = 3$

5.1.3 Displaying the computed variograms

Once the parameters are entered and *Next* is clicked, SGeMS computes and displays the variograms (see Fig. 5.3). There is one plot per requested direction, plus an additional plot showing all directions together (upper left plot on Fig. 5.3). A right-click on a plot will toggle the display of the number of pairs used to compute each variogram value. Clicking on the square icon at the top of a plot will "maximize" the plot, i.e. the plot will occupy the maximum space, hiding the other plots. In maximized mode, use the *Ctrl + Tab* key combination to toggle between the plots. If a plot is closed by clicking on its cross icon, it can not be re-opened without re-computing the variograms.

Above the plots, three menus provide means to re-arrange the plots, modify their scale, and save them as images or as text.

> **File → Save Experimental Variograms** Save the computed experimental variogram values to a file. That file can be loaded later from the first screen of the variogram tool (see item **1**, p. 93).
>
> **File → Export Plots As Images** Save the plots to image files. It is possible to select which plots will be saved.
>
> **Edit → Plot Settings** Modify the axis scale of all or some plots.
>
> **Window → Tile** Arrange the plots so that they occupy all the available space.
>
> **Window → Cascade** Display the plots in cascade.

5.2 Variogram modeling in SGeMS

SGeMS provides an interface to interactively fit a variogram model of the form

$$\gamma(\mathbf{h}) = c_0\gamma_0(\mathbf{h}) + \sum_{n=1}^{N} c_n\gamma_n(\mathbf{h}) \tag{5.5}$$

to the computed experimental variograms. In Eq. (5.5) γ_0 is a pure nugget effect, γ_n ($n > 0$) is either a spherical, exponential or Gaussian variogram model, and $c_0, \dots, c_N$ are the variance contributions of each of the nested structures $\gamma_0, \dots, \gamma_N$.

Notice that while SGeMS can compute other correlation measures than variograms such as covariances and correlograms, it can only model variograms.

The variogram modeling interface is shown in Fig. 5.3 and is accessed after either computing an experimental variogram or loading an existing experimental variogram (Section 5.1). The right-hand side panel allows to enter a variogram model of form Eq. (5.5) which will be overlaid on the experimental variogram plots. The parameters of the model can then be interactively modified to fit the experimental variogram.

Input of a variogram model

The interface used to input a variogram model is shown on Fig. 5.3, right-hand side panel.

Description of the interface

19. **Nugget Effect** The contribution to the sill of the nugget effect, c_0 in Eq. (5.5).
20. **Nb of structures** The number N of nested structures.
21. **Sill Contribution** Contribution to the sill of the nth structure, c_n in Eq. (5.5).
22. **Type** The type of variogram for that structure. The three possible types are: spherical, exponential and Gaussian.
23. **Ranges** The ranges of the variogram. Ranges can either be changed manually by entering the value, or by dragging the corresponding *slider*. A *slider* allows to continuously change the range value, between 0 and a fixed maximum. If the desired range is greater than the preset maximum of the slider, it must be input in the text field. The maximum of the slider will be increased accordingly. Use the *Reset* button to reset the maximum of the slider to its default.
24. **Angles** The angles defining the variogram model anisotropy ellipsoid. The first angle is the azimuth, the second the dip, and the third the rake, see Section 2.5. All angles must be entered in degrees. For 2D modeling, the dip and rake should be set as 0.

Saving the model

Once a model has been fit to the experimental variogram, it can be saved to a file using the *File → Save Variogram Model* menu item. That file can then be used to specify a variogram model to geostatistical algorithms.

Modeling a coregionalization

Modeling the coregionalization of two random functions $Z_1(\mathbf{u})$ and $Z_2(\mathbf{u})$ calls for the computation and joint modeling of four experimental variograms $\hat{\gamma}_{1,1}, \hat{\gamma}_{1,2}, \hat{\gamma}_{2,1}, \hat{\gamma}_{2,2}$. The four experimental variograms can not be modeled independently from one another since the variogram matrix

$$\Gamma = \begin{bmatrix} \gamma_{1,1} & \gamma_{1,2} \\ \gamma_{2,1} & \gamma_{2,2} \end{bmatrix}$$

must be conditionally negative definite ($\gamma_{i,j}$ models $\hat{\gamma}_{i,j}$).

SGeMS does not provide specific tools to model a coregionalization. Each experimental (cross-)variogram would have to be computed and modeled within its own variogram modeling window. It is then the responsibility of the user to ensure that the final model Γ is a permissible model; see Goovaerts (1997, p.117) on how to fit a linear model of coregionalization.

Note that many SGeMS algorithms support models of coregionalizations, such as the Markov Models 1 and 2, that alleviate the need to jointly model all four variograms $\hat{\gamma}_{1,1}$, $\hat{\gamma}_{1,2}$, $\hat{\gamma}_{2,1}$, $\hat{\gamma}_{2,2}$ (see Section 3.6.4 for model details).

6

Common parameter input interfaces

Parameters to the SGeMS algorithms are usually input through their graphical interfaces.[1] Although each algorithm has its own specific interface, all share standard elements, for instance to select a grid, a property, or parametrize a variogram or a distribution. The purpose of this chapter is to describe how to use these recurring graphical elements.

6.1 Algorithm panel

When an algorithm is selected from the algorithms panel, the corresponding parameters graphical interface is displayed (see Fig. 2.2).

The algorithm panel, briefly described in Section 2.1, is shown in Fig. 6.1. The main interface has six parts.

Parameters description

1. **Algorithms**   List of all available algorithms which are grouped into three classes: Estimation, Simulation and Utilities.
2. **Parameters input**   The graphical parameter interface. The parameters for the selected algorithm are entered in this area.
3. *Parameters → Load*   Load parameters previously saved in a file. The parameters can also be loaded by dragging the parameter file into the graphical parameter interface window.
4. *Parameters → Save*   Save the parameters already entered in the graphical interface to a file. It is *recommended* that the parameter file has the extension ".par", and should be saved outside of the SGeMS project folder.
5. *Parameters → Clear All*   Clear all parameters entered in the current interface.

---

[1] Chapter 10 explains how to launch the algorithms without going through the graphical interface

Common parameter input interfaces

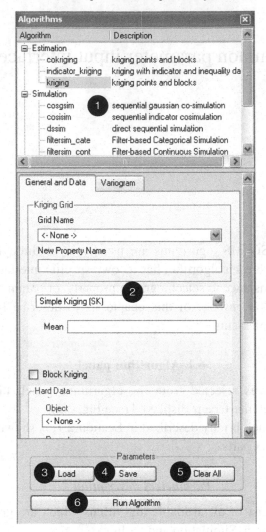

Figure 6.1 Algorithm panel

6. Run Algorithm   Run the selected algorithm with the entered parameters.

6.2 Selecting a grid and property

Selecting a property is done in all algorithm user interfaces for tasks such as, but not limited to, choosing the conditioning data or a training image. It is done through the property selector interface, see Fig. 6.2. That selector is linked to a grid chooser: once a grid is chosen, a list of all the properties belonging to the grid is displayed and the user can select the appropriate property.

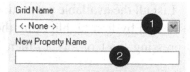

Figure 6.2 Single property selection widget

Parameters description

1. **Grid Name**   Select an object from the list of all available objects currently loaded in SGeMS. The object can either be a Cartesian grid or a point-set. Only one object can be selected.
2. **Property Name**   Select a property from the property list of the selected object. Only one property can be highlighted and selected.

6.3 Selecting multiple properties

The selection of multiple properties is done with the multiple property selector, see Fig 6.3, which also allows to order the selected properties. Ordering properties is often necessary, for instance when properties are associated with an ordered set of thresholds or categories. In that case, the first property must relate to the first category, the second property to the second category and so on.

Parameters description

1. **Selected Properties**   The selected properties will appear in this window.
2. **Choose Properties**   Button to select the properties. A selection window will pop up, as shown on the right hand side of Fig. 6.3.

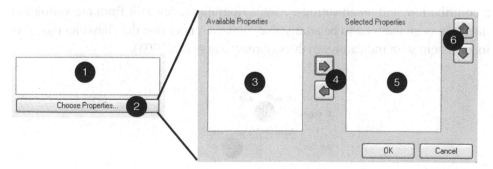

Figure 6.3 Multiple properties selection interface

3. **Available Properties**   List all the available properties in the current working object. Properties are selected by first highlighting them (multiple properties can be highlighted by pressing `Ctrl` or `Shift`) then clicking the right arrow button, see item **4**.

4. **Properties selector**   Use arrows to move properties back and forth between the available property (item **3**) and the selected property window (see hereafter item **5**). Only highlighted properties are transferred.

5. **Selected Properties**   List of currently selected properties. Those properties can be unselected with the left arrow in item **4**.

6. **Properties ordering**   Order the selected properties. The top property is the first one and the bottom property is the last one. To change the order, highlight the property first, then use the up or down arrows to correct the sequence.

6.4 Search neighborhood

Figure 6.4 shows the search neighborhood interface, which parametrizes an ellipsoid by its maximum, medium and minimum axes. These axes are positioned in space through three angles, see Section 2.5.

Parameters description

1. **Ranges**   Maximum, medium and minimum ranges of the search ellipsoid.
2. **Angles**   Rotation angles for anisotropic ellipsoid.

6.5 Variogram

The variogram interface is used in all variogram-based algorithms in SGeMS, see Fig. 6.5. This interface allows the specification of variograms with nested structures. Each nested structure is independently parametrized by a variogram type, a contribution, and an anisotropy. Any variogram model built from the variogram interface is guaranteed to be admissible; however, note that the Gaussian model is inconsistent with indicator variables (Armstrong *et al.*, 2003).

Figure 6.4 Search ellipsoid interface

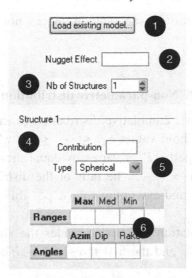

Figure 6.5 Variogram interface

Parameters description

1. **Load existing model**   Initialize the variogram from a saved variogram file.
2. **Nugget effect**   Value for the nugget effect.
3. **Nb of structures**   Number of nested structures, excluding the nugget effect. For n structures, the following items **4** to **6** will be repeated n times.
4. **Contribution**   Sill for the current structure.
5. **Type**   Type of variogram for the selected structure (spherical, exponential or Gaussian).
6. **Anisotropy**   Maximum, medium and minimum ranges and rotation angles. In 2D, the dip and rake rotations should be 0 and the minimum range must be less than the medium range.

6.6 Kriging

The selection of the kriging type is done with a special interface. The available kriging types are simple kriging (SK), ordinary kriging (OK), kriging with a trend (KT) and kriging with a local varying mean (LVM). Only OK does not require extra parameters; SK requires a mean, KT requires the components of the polynomial trend (Section 3.6.2) and LVM requires the property within which the local means are stored.

6.7 Line entry

The line entry interface is often used either to enter a name, for instance the name of a new property to be created, or to enter a series of values such a thresholds. Note that any numerical series must be separated by spaces, not commas or semicolons. The entry is case sensitive.

6.8 Non-parametric distribution

In SGeMS a non-parametric cumulative distribution function, cdf $F(z)$, is determined from a set of threshold values $z_1 \leq \cdots \leq z_L$ which can either be read from a file or from a property. $F(z)$ varies by equal increment $1/(L+1)$ with: $F(z_1) = \frac{1}{L+1}$ and $F(z_L) = \frac{L}{L+1}$. The tails of the distribution are obtained by extrapolating to minimum and maximum values, possibly less than z_1 and greater than z_L respectively.

The lower tail extrapolation function provides the shape of the distribution between the minimum z_{min} and the first threshold z_1. The options for the lower tail are as follows.

- Z is bounded: $F(z_{min}) = 0$. The lower tail of F is then modeled with a power model:

$$\frac{F(z_1) - F(z)}{F(z_1)} = \left(\frac{z_1 - z}{z_1 - z_{min}}\right)^{\omega} \quad \forall z \in (z_{min}, z_1). \tag{6.1}$$

Parameter ω controls the decrease of the function, with the constraint $\omega \geq 1$. The greater ω the less likely are low values close to z_{min}. For $\omega = 1$, all values between z_{min} and z_1 are equi-probable.

- Z is not bounded: the lower tail is modeled with an exponential function:

$$F(z) = F(z_1) \exp\left(-(z - z_1)^2\right) \quad \forall z < z_1. \tag{6.2}$$

The options for the upper tail extrapolation function are similar but applied to the interval (z_L, z_{max}).

- Z is bounded: $F(z_{max}) = 1$. The upper tail of F is then modeled with a power model:

$$\frac{F(z) - F(z_L)}{1 - F(z_L)} = \left(\frac{z - z_L}{z_{max} - z_L}\right)^{\omega} \quad \forall z \in (z_L, z_{max}). \tag{6.3}$$

Parameter ω controls the decrease of the function, with the constraint $\omega \in [0, 1]$. The lower the ω value the less likely are extreme values close to z_{max}. For $\omega = 1$, all values between z_L and z_{max} are equi-probable.

- Z is not bounded: the upper tail is modeled by an hyperbolic model:

$$\frac{1 - F(z)}{1 - F(z_L)} = \left(\frac{z_L}{z}\right)^{\omega} \quad \forall z > z_L, \ \omega \geq 1. \tag{6.4}$$

All $L - 1$ intermediary intervals $[z_i, z_{i+1}]$ for $i = 1, ..., L - 1$ are interpolated linearly, corresponding to a power model with $\omega = 1$.

Note: when z_{min} and z_{max} values are set to z_1 and z_L, there is no need for tail extrapolation.

Tie breaking

SGeMS allows to randomly break ties within a non-parametric distribution z_i, $i = 1, ..., L$. Consider the case where n of the L values are identical: $z_i = z_{i+1} = \cdots = z_{i+n}$, $i + n < L$. Instead of assigning the same cdf value of $F(z_{i+n})$ to the n data $z_i, ..., z_{i+n}$, the cdf values $F(z_i), ..., F(z_{i+n})$ are randomly assigned the values $i/(L + 1), ..., (i + n)/(L + 1)$. This is analogous of adding a very small noise to each tie value.

Parameters description

The non-parametric distribution interface is shown in Fig. 6.6, and the parameters are described below.

1. **Reference distribution** Read the reference distribution data either from a data file [ref_on_file] or from a grid [ref_on_grid].
2. **Break ties** [break_ties] Randomly break tied values when assigning their corresponding cdf values. There will be as many different cdf values as there are distribution values.

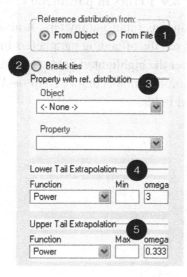

Figure 6.6 Interface for non-parametric distribution

3. **Source for reference distribution**   If [ref_on_grid] is selected, the distribution values are recorded in a currently loaded SGeMS property. [grid] and [property] contain the values for the non-parametric distributions. If [ref_on_file] is selected, the input data file containing the reference distribution is entered in [filename]. The reference distribution must be given in one column without header with numbers only.

4. **Lower Tail Extrapolation**   Parametrization of the lower tail. The type of extrapolation function is selected with [LTI_function]. If the power model is selected, the minimum value z_{min} [LTI_min] and the parameter ω [LTI_omega] must be specified. Note that the minimum [LTI_min] must be less than or equal to the minimum datum value as entered in the reference distribution, and the power ω [LTI_omega] must be greater or equal to 1. The exponential model does not require any parameter. No parameters are required when no extrapolation is required.

5. **Upper Tail Extrapolation**   Parametrization of the upper tail. The type of extrapolation function is selected with [UTI_function]. If the power model is selected, the maximum value z_{max} [UTI_max] and the parameter ω [UTI_omega] must be specified. Note that the maximum [UTI_max] must be greater than or equal to the maximum datum value as entered in the reference distribution, and the power ω [UTI_omega] must be less than or equal to 1. The hyperbolic model only requires parameter ω [UTI_omega] when the upper tail is unbounded. No parameters are required when no extrapolation is required.

6.9 Errors in parameters

When SGeMS detects erroneous input parameters, it aborts the execution of the algorithm and highlights in red the offending parameters in the algorithm interface. Leave the mouse pointer over the highlighted fields to get a description of the error, or alternatively, select the question mark cursor (*Help → What's this* or Shift-F1) and click on the highlighted fields.

7

Estimation algorithms

This chapter presents the SGeMS collection of estimation algorithms related to the kriging estimation formalism. First the algorithm *KRIGING* is presented. It performs estimation of a single variable either by simple kriging, ordinary kriging, kriging with a local varying mean (LVM) or kriging with a trend. *KRIGING* also allows the possibility to estimate block values from point data. Next, the *COKRIG-ING* algorithm is presented. The information carried by a secondary variable can be integrated with a Markov Model 1 or 2 (MM1 or MM2) or with a linear model of coregionalization (LMC). The third estimation algorithm is the non-parametric *INDICATOR KRIGING* (IK), which consists of simple kriging applied to binary indicator data. The last estimation algorithm is *BKRIG*, kriging with linear average variable, which allows to estimate property with point and/or block support data.

All these estimation algorithms require a search neighborhood within which to select the relevant neighboring data. The estimation procedure is only carried forward if a minimum number of conditioning data is found. Otherwise, the central node is left not informed and a warning message is issued. In such case the user could increase the search neighborhood to allow more data to be considered.

7.1 *KRIGING*: univariate kriging

Kriging is a generalized regression method that provides the best estimate in the least square sense, see Section 3.6. SGeMS can build and solve four types of kriging systems depending on the stationarity assumption about the mean of the random function model.

Simple kriging (SK) The mean of the domain is assumed constant and known.
Ordinary kriging (OK) The mean inside each estimation neighborhood is unknown but constant; it is evaluated by the OK algorithm from the neighborhood data.

Kriging with a trend (KT) The mean follows a functional trend $m(\mathbf{u}) = f(x, y, z)$. SGeMS allows a polynomial trend, either linear or quadratic in any of the three coordinates x, y, z.

Kriging with a local varying mean (LVM) The mean varies from location to location and is given as a secondary data.

The kriging algorithm is given in Algorithm 7.1.

Algorithm 7.1 General kriging

1: **for** Each location $\mathbf{u}$ in the grid **do**
2:    Get the conditioning data $n(\mathbf{u})$
3:    **if** $n(\mathbf{u})$ is large enough **then**
4:       Build the kriging system from the $n(\mathbf{u})$ neighboring data and solve it
5:       Compute the kriging estimate and the kriging variance for location $\mathbf{u}$
6:    **else**
7:       Set node as uninformed
8:    **end if**
9: **end for**

For block kriging, only the SK and OK options are available. The regularized variogram between point and block is internally computed by discretizing the block into a user-defined number of points in the x, y, and z directions. For instance, the average point-support variogram between a point $\mathbf{u}$ and the discretizing points inside a block $V(\mathbf{u})$, see Section 3.6.3 and Fig. 7.1, is approximated with:

$$\gamma_V(\mathbf{u}, V(\mathbf{u})) = \frac{1}{M} \sum_{\mathbf{u}'=1}^{M} \gamma(\mathbf{u}, \mathbf{u}'),$$

where M is the number of points discretizing the block $V(\mathbf{u})$.

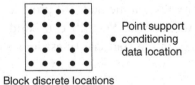

Point support
● conditioning
data location

Block discrete locations

Figure 7.1 Variogram averaging procedure used for block kriging. The variogram value between the point support conditioning data and the unknown block value is the average variogram between the conditioning data and all the points discretizing the block

Parameters description

The *KRIGING* algorithm is activated from *Estimation* → *kriging* in the algorithm panel. The *KRIGING* interface contains two pages: "General and Data" and "Variogram" (see Fig. 7.2). The text inside "[ ]" is the corresponding keyword in the *KRIGING* parameter file.

1. **Grid Name** [Grid_Name]   Name of the estimation grid.
2. **New Property Name** [Property_Name]   Name of the kriging output. A second property with the suffix _krig_var is also created to store the kriging variance.
3. **Kriging Type** [Kriging_Type]   Select the type of kriging system to be solved at each node.
4. **Block kriging option** [do_block_kriging] If [do_block_kriging] is selected then the X, Y and Z block discritization is given in [npoints_X], [npoints_Y] and [npoints_Z]. When the block kriging option is selected, the conditioning data must be given on a point set. Note that the LVM and KT options are not available for block kriging.

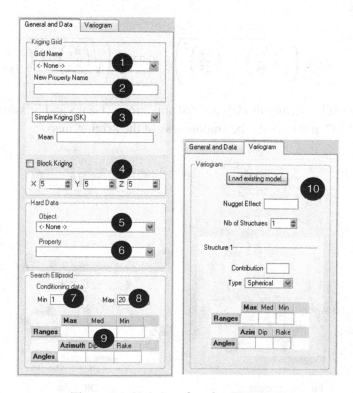

Figure 7.2 User interface for *KRIGING*

5. **Hard Data—Object** [Hard_Data.grid] Name of the grid containing the conditioning data.

6. **Hard Data—Property** [Hard_Data.property] Property for the conditioning data.

7. **Min Conditioning data** [Min_Conditioning_Data] Minimum number of data to be retained in the search neighborhood.

8. **Max Conditioning data** [Max_Conditioning_Data] Maximum number of data to be retained in the search neighborhood.

9. **Search Ellipsoid Geometry** [Search_Ellipsoid] Parametrization of the search ellipsoid, see Section 6.4.

10. **Variogram** [Variogram] Parametrization of the variogram, see Section 6.5.

Example

Ordinary kriging is performed on the Ely1 data set (described in Section 4.1.1) using the 50 hard data; the results are shown in Fig. 7.3 along with the corresponding ordinary kriging variance. The search neighborhood is isotropic with radius 120, the neighborhood must contain at least 5 data but no more than 25. The model variogram is

$$\gamma(h_x, h_y) = 1.2 Sph\left(\sqrt{\left(\frac{h_x}{35}\right)^2 + \left(\frac{h_y}{45}\right)^2}\right) + 0.2 Sph\left(\sqrt{\left(\frac{h_x}{35}\right)^2 + \left(\frac{h_y}{100000}\right)^2}\right),$$

where the second structure models a zonal anisotropy (Isaaks and Srivastava, 1989; Goovaerts, 1997, p.93). Note the smoothness of the kriging map.

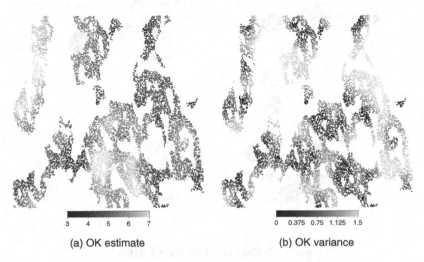

<table>
<tr><td>3</td><td>4</td><td>5</td><td>6</td><td>7</td></tr>
</table>

<table>
<tr><td>0</td><td>0.375</td><td>0.75</td><td>1.125</td><td>1.5</td></tr>
</table>

(a) OK estimate   (b) OK variance

Figure 7.3 Kriging estimate and variance with ordinary kriging

7.2 *INDICATOR KRIGING*

INDICATOR KRIGING is a non-parametric estimation algorithm used to estimate the conditional cumulative distribution function at any location given neighboring conditioning data. This algorithm can be used with both categorical and continuous variables, see Section 3.6 for a brief recall of the theory of indicator kriging. Indicator kriging does not guarantee that the resulting distributions are valid, i.e. that the probabilities monotonically increase and are bound by 0 and 1 for the continuous case, or that they are all positive summing to 1 for the categorical case. When those requirements are not met, an order correction is performed by the algorithm (Deutsch and Journel, 1998, p.81).

Continuous variable

Continuous indicator variables are defined as:

$$i(\mathbf{u}, z_k) = \begin{cases} 1 & \text{if } z(\mathbf{u}) \leq z_k \\ 0 & \text{otherwise.} \end{cases}$$

The aim of indicator kriging is to estimate the probability of $Z(\mathbf{u})$ being less than threshold value z_k, conditional to the data (n) retained:

$$I^*(\mathbf{u}, z_k) = E^*(I(\mathbf{u}, z_k) \mid (n))$$
$$= \text{Prob}^*(Z(\mathbf{u}) < z_k \mid (n)). \tag{7.1}$$

Estimating $I^*(\mathbf{u}, z_k)$ for different cutoffs z_k, $k = 1, \ldots, K$, yields a discrete estimate of the conditional cumulative distribution function (ccdf) of $Z(\mathbf{u})$ at threshold values $z_1, \ldots, z_K$.

The estimated value $i^*(\mathbf{u}; z_k)$ at location $\mathbf{u}$, is seen as an estimate of $\text{Prob}^*(Z(\mathbf{u}) \leq z_k|(n))$ (Goovaerts, 1997, p.293). The complete ccdf $F(z|(n))$ is then reconstituted and any conditional statistics can be retrieved.

Algorithm *INDICATOR KRIGING* estimates the conditional probabilities $i^*(\mathbf{u}; z_k)$ by solving a simple kriging system, hence assuming that the marginal probabilities $E\{I(\mathbf{u}, z_k)\}$ are known and constant (independent of $\mathbf{u}$). No ordinary indicator kriging option is available in this initial version of SGeMS. Two types of regionalization models are considered for indicator kriging. The full IK option requires a variogram model for each threshold, see Algorithm 7.2. The median IK (Goovaerts, 1997, p.304) option only requires the variogram model for the median threshold, all other indicator variograms are then assumed proportional to that single model, see Algorithm 7.3.

Coding information as indicator values

Consider a continuous variable $Z(\mathbf{u})$ and a discretization of its range by the K threshold values $z_1, \ldots, z_K$.

Algorithm 7.2 Full indicator kriging

1: **for** Each location **u** in the grid **do**
2:   **for** Each category k **do**
3:     Get the conditioning data n
4:     **if** n is large enough **then**
5:       Solve simple indicator kriging system
6:       Compute the kriging estimate $i_k^*(\mathbf{u})$
7:     **else**
8:       Set node as un-informed, move to the next location
9:     **end if**
10:   **end for**
11:   Correct $F_Z(\mathbf{u})$ for order relation violations
12: **end for**

Algorithm 7.3 Median indicator kriging

1: **for** Each location **u** in the grid **do**
2:   Get the conditioning data n
3:   **if** n is large enough **then**
4:     Solve simple indicator kriging system and store the vector of kriging weights
5:   **else**
6:     Set node as uninformed, move to the next location
7:   **end if**
8:   **for** Each category k **do**
9:     Compute kriging estimate $i_k^*(\mathbf{u})$ with the kriging weights found in step 4
10:   **end for**
11:   Correct $F_Z(\mathbf{u})$ for order violations
12: **end for**

Different types of data can be coded into a vector of K indicator values $\mathbf{I}(\mathbf{u}) = [i(\mathbf{u}; z_1), \ldots, i(\mathbf{u}; z_K)]$.

Hard data

The value of Z at a given location $\mathbf{u}_\alpha$ is known, equal to $z(\mathbf{u}_\alpha)$ with no uncertainty. The corresponding K indicator values are all valued 0 or 1:

$$
i(\mathbf{u}_\alpha; z_k) = \begin{cases} 1 & \text{if } z(\mathbf{u}_\alpha) \leq z_k \\ 0 & \text{otherwise} \end{cases} \quad k = 1, \ldots, K.
$$

Example If there are $K = 5$ threshold values: $z_1 = 1$, $z_2 = 2$, $z_3 = 3$, $z_4 = 4$, $z_5 = 5$, then a hard datum value $z = 3.7$ would be coded as the following vector of 5 indicator data:

$$
I = \begin{bmatrix} 0 \\ 0 \\ 0 \\ 1 \\ 1 \end{bmatrix}.
$$

Inequality data

The datum value $z(\mathbf{u}_\alpha)$ is known only to lie within an interval, e.g. $z(\mathbf{u}_\alpha) \in [a, b]$, or $z(\mathbf{u}_\alpha) \in]a, +\infty[$. The indicator data vector $I(\mathbf{u}_\alpha)$ from the information $z(\mathbf{u}_\alpha) \in [a, b[$ is:

$$
i(\mathbf{u}_\alpha; z_k) = \begin{cases} 0 & \text{if } z_k < a \\ \text{missing} & \text{if } z_k \in [a, b[\\ 1 & \text{if } z_k \geq b \end{cases} \qquad k = 1, \ldots, K.
$$

Example If $K = 5$, with $z_1 = 1$, $z_2 = 2$, $z_3 = 3$, $z_4 = 4$, $z_5 = 5$, then the interval datum $[2.1, 4.9]$ is coded as:

$$
I = \begin{bmatrix} 0 \\ 0 \\ ? \\ ? \\ 1 \end{bmatrix}
$$

where the question mark ? denotes an undefined value (missing values are represented by integer -9966699 in SGeMS) which would have to be estimated by kriging. See Section 2.2.7 on how to enter missing values in SGeMS.

Categorical variable

INDICATOR KRIGING can be applied to categorical variables, i.e. variables that take a finite number K of discrete values (also called classes or categories): $z(\mathbf{u}) \in \{0, \ldots, K - 1\}$. The indicator variable for class k is defined as:

$$
I(\mathbf{u}, k) = \begin{cases} 1 & \text{if } Z(\mathbf{u}) = k \\ 0 & \text{otherwise} \end{cases}
$$

and the probability $I^*(\mathbf{u}, k)$ for $Z(\mathbf{u})$ belonging to class k is estimated by simple kriging:

$$
I^*(\mathbf{u}, k) - E\{I(\mathbf{u}, k)\} = \sum_{\alpha=1}^{n} \lambda_\alpha(\mathbf{u})\Big(I(\mathbf{u}_\alpha, k) - E\{I(\mathbf{u}_\alpha, k)\}\Big)
$$

where $E\{I(\mathbf{u}, k)\}$ is the indicator mean (marginal probability) for class k.

In the case of categorical variables, the estimated probabilities must all be in $[0, 1]$ and verify:

$$\sum_{k=1}^{K} I^*(\mathbf{u}; k) = 1. \tag{7.2}$$

If not, they are corrected as follows:

1. If $I^*(\mathbf{u}, k) \notin [0, 1]$ reset it to the closest bound. If all the probability values are less than or equal to 0, no correction is made and a warning is issued.
2. Standardize the values so that they sum-up to 1:

$$I^*_{\text{corrected}}(\mathbf{u}, k) = \frac{I^*(\mathbf{u}, k)}{\sum_{i=1}^{K} I^*(\mathbf{u}, i)}.$$

Parameters description

The *INDICATOR KRIGING* algorithm is activated from *Estimation* → Indicator Kriging in the algorithm panel. The *INDICATOR KRIGING* interface contains three pages: "General", "Data" and "Variogram" (see Fig. 7.4). The text inside "[ ]" is the corresponding keyword in the *INDICATOR KRIGING* parameter file.

1. **Estimation Grid Name** [Grid_Name]   Name of the estimation grid.
2. **Property Name Prefix** [Property_Name]   Prefix for the estimation output. The suffix _real# is added for each indicator.
3. **# of indicators** [Nb_Indicators]   Number of indicators to be estimated.

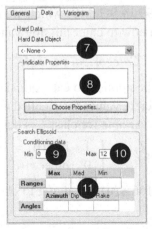

Figure 7.4 User interface for *INDICATOR KRIGING*

4. **Categorical variable** [Categorical_Variable_Flag]   Indicates if the data are categorical or not.

5. **Marginal probabilities** [Marginal_Probabilities]

   If continuous Probability to be below the thresholds. There must be [Nb_Indicators] entries monotonically increasing.

   If categorical Proportion for each category. There must be [Nb_Indicators] entries adding to 1. The first entry corresponds to category coded 0, the second to category coded 1, ...

6. **Indicator kriging type**   If **Median IK** [Median_Ik_Flag] is selected, the program uses median indicator kriging to estimate the ccdf. Otherwise, if **Full IK** [Full_Ik_Flag] is selected, a different IK system is solved for each threshold/class.

7. **Hard Data Grid** [Hard_Data_Grid]   Grid containing the conditioning hard data.

8. **Hard Data Indicators Properties** [Hard_Data_Property]   Conditioning primary data for the simulation. There must be [Nb_Indicators] properties selected, the first one being for class 0, the second for class 1, and so on. If **Full IK** [Full_Ik_Flag] is selected, a location may not be informed for all thresholds.

9. **Min Conditioning data** [Min_Conditioning_Data]   Minimum number of data to be retained in the search neighborhood.

10. **Max Conditioning data** [Max_Conditioning_Data]   Maximum number of data to be retained in the search neighborhood.

11. **Search Ellipsoid Geometry** [Search_Ellipsoid]   Parametrization of the search ellipsoid, see Section 6.4.

12. **Variogram** [Variogram]   Parametrization of the indicator variograms, see Section 6.5. Only one variogram is necessary if **Median IK** [Median_Ik_Flag] is selected. Otherwise there are [Nb_Indicators] indicator variograms.

Example

The *INDICATOR KRIGING* algorithm is run on the point-set presented in Fig. 4.1a. Probabilities of having a value below 4, 5.5 and 7 are computed with a median IK regionalization. The resulting conditional probabilities (ccdf) for these three thresholds are: 0.15, 0.5 and 0.88. The estimated probabilities for each threshold are shown in Fig. 7.5. The variogram model for the median indicator is:

$$\gamma(h_x, h_y) = 0.07 Sph\left(\sqrt{\left(\frac{h_x}{10}\right)^2 + \left(\frac{h_y}{15}\right)^2}\right) + 0.14 Sph\left(\sqrt{\left(\frac{h_x}{40}\right)^2 + \left(\frac{h_y}{75}\right)^2}\right).$$

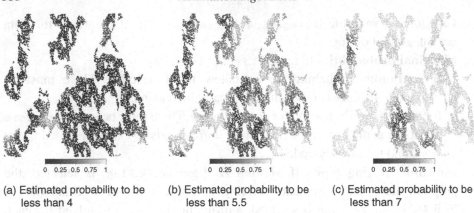

(a) Estimated probability to be   (b) Estimated probability to be   (c) Estimated probability to be
     less than 4   less than 5.5   less than 7

Figure 7.5 Median indicator kriging

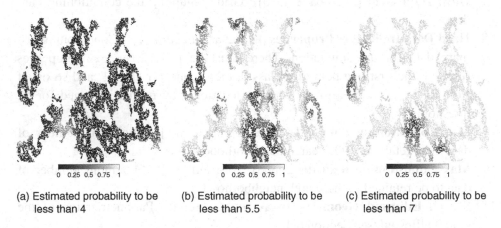

(a) Estimated probability to be   (b) Estimated probability to be   (c) Estimated probability to be
     less than 4   less than 5.5   less than 7

Figure 7.6 Median indicator kriging with inequality data

The search ellipsoid is of size $80 \times 80 \times 1$, with a minimum of 5 and maximum of 25 conditioning data.

A set of 200 interval-type data is added to the data set, these interval data only tell whether the z-values at these locations are above or below 5.5. The coding of these inequality data is:

$$\text{if } z(\mathbf{u}) < 5.5 \text{ then } i(\mathbf{u}) = \begin{bmatrix} ? \\ ? \\ 1 \end{bmatrix} ; \quad \text{if } z(\mathbf{u}) > 5.5 \text{ then } i(\mathbf{u}) = \begin{bmatrix} 0 \\ 0 \\ ? \end{bmatrix}.$$

Figure 7.6 shows the resulting estimation maps for the same thresholds used in Fig. 7.5. The inequality data mostly modify the probability map for the 5.5 threshold, their impact on the low and high threshold is not as strong.

7.3 *COKRIGING*: kriging with secondary data

The *COKRIGING* algorithm integrates the information carried by a secondary variable related to the primary attribute being estimated. The kriging system of equations is then extended to take into account that extra information. A coregionalization model must be provided to integrate secondary variables. SGeMS offers three choices: the linear model of coregionalization (LMC), the Markov Model 1 (MM1) and the Markov Model 2 (MM2). The LMC accounts for all secondary data within the search neighborhood while the Markov models only retain those secondary data that are co-located with the primary data; see Section 3.6.4.

The LMC option can be used with either simple or ordinary cokriging. The Markov models (MM1 or MM2) can only be solved with simple cokriging; using ordinary cokriging would lead to ignoring the secondary variable since the sum of weights for the secondary variable must be equal to zero (Goovaerts, 1997, p.236).

The detailed *COKRIGING* algorithm is presented in Algorithm 7.4.

Algorithm 7.4 *COKRIGING*

1: **for** Each location **u** in the grid **do**
2:      Get the primary conditioning data n
3:      Get the secondary conditioning data n'
4:      **if** n is large enough **then**
5:          Solve the cokriging system
6:          Compute the cokriging estimate and cokriging variance
7:      **else**
8:          Set node as uninformed, move to the next location
9:      **end if**
10: **end for**

> **Tip 1** Joint-estimation of several attributes
> *The SGeMS cokriging algorithm only allows for estimation of a primary attribute conditional to primary data and data coming from a single secondary attribute. There are applications where several (>2) correlated attributes need to be jointly estimated. Such estimation can be performed in SGeMS by first orthogonalizing all primary and secondary attributes into factors using some type of principal components factorization (Vargas-Guzman and Dimitrakopoulos, 2003), modeling the variogram for each of the factors independently, performing kriging on each factor, then transforming them back into estimates of the original attributes. With the embedded Python scripting ability, all the previous steps can be performed within SGeMS (principal component decomposition can be performed using the scipy Python library).*

Parameters description

The *COKRIGING* algorithm is activated from *Estimation | cokriging* in the Algorithm Panel. The main *COKRIGING* interface contains three pages: "General", "Data" and "Variogram" (see Fig. 7.7). The text inside "[ ]" is the corresponding keyword in the *COKRIGING* parameter file.

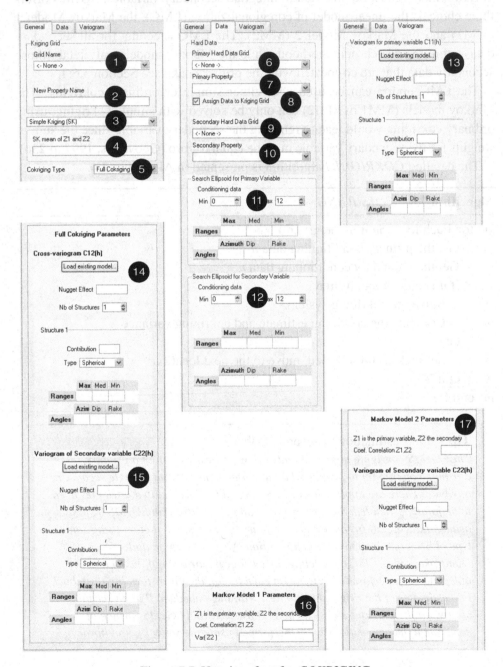

Figure 7.7 User interface for *COKRIGING*

1. **Grid Name** [Grid_Name]   Name of the estimation grid.

2. **New Property Name** [Property_Name]   Name of the cokriging output. A second property with the suffix _krig_var is also created.

3. **Kriging Type** [Kriging_Type]   Select the type of kriging system to be solved at each node.

4. **SK mean of Z1 and Z2** [SK_Means]   Means of the primary and secondary data. Only required if [Kriging_Type] is set to simple kriging.

5. **Cokriging Type** [Cokriging_Type]   Model of coregionalization used to integrate the secondary information. Note that MM1 and MM2 models cannot be used with ordinary cokriging.

6. **Primary Hard Data—Object** [Hard_Data.grid]   Name of the grid containing the conditioning data.

7. **Primary Hard Data—Property** [Hard_Data.property]   Property for the conditioning data.

8. **Assign Hard Data to Grid** [Assign_Hard_Data]   If selected, the hard data are copied on the estimation grid. The program does not proceed if the copying fails. This option significantly increases execution speed.

9. **Secondary Hard Data—Object** [Hard_Data.grid]   Name of the grid containing the secondary variable conditioning data.

10. **Secondary Hard Data—Property** [Hard_Data.property]   Property for the conditioning data.

11. **Search Ellipsoid for Primary Variable**   The minimum and maximum number of primary data to be retained in the search neighborhood are given in [Min_Conditioning_Data] and [Max_Conditioning_Data]. The search ellipsoid geometry is parametrized with [Search_Ellipsoid_1], see Section 6.4.

12. **Search Ellipsoid for Secondary Variable**   The minimum and maximum number of secondary data to be retained in the search neighborhood are given in [Min_Conditioning_Data_2] and [Max_Conditioning_Data_2]. The search ellipsoid geometry is parametrized with [Search_Ellipsoid_2]. These entries are only required when [Cokriging_Type] is set to **Full Cokriging**.

13. **Variogram for primary variable** [Variogram_C11]   Parametrization of the variogram for the primary variable, see Section 6.5.

14. **Cross-variogram between primary and secondary variables** [Variogram_C12]   Parametrization of the cross-variogram between the primary and secondary variables. Required if **Cokriging Option** [Cokriging_Type] is set to **Full Cokriging**.

15. **Variogram for secondary variable** [Variogram_C22]   Parametrization of the variogram of the secondary variable. Required if **Cokriging Option** [Cokriging_Type] is set to **Full Cokriging**.

16. **MM1 parameters**   The coefficient of correlation between the primary and secondary variables is entered in [Correl_Z1Z2]. The variance of the

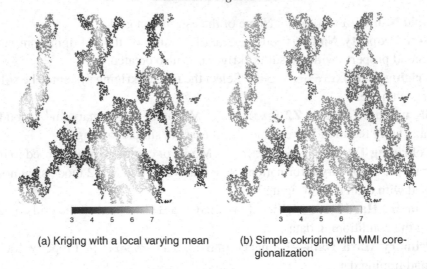

(a) Kriging with a local varying mean   (b) Simple cokriging with MMI core-
                                                   gionalization

Figure 7.8 Kriging with additional information: (a) the local mean and (b) a correlated secondary attribute

secondary attribute is given by [Var_Z2] Only required if **Cokriging Option** [Cokriging_Type] is set to **MM1**.

17. **MM2 parameters**   The coefficient of correlation between the primary and secondary variables is entered in [MM2_Correl_Z1Z2]. The variogram of the secondary attribute is given by [MM2_Variogram_C22]. Only required if **Cokriging Option** [Cokriging_Type] is set to **MM2**.

Example

Simple cokriging with the MM1 model is used to estimate the Ely1 point-set with the exhaustive secondary variable data shown in Fig. 4.1. The parameters related to the primary attribute (minimum and maximum conditioning data, search neighborhood and variogram) are the same as with the kriging example shown in Fig. 7.3. The MM1 model is parametrized with a coefficient of correlation 0.71 between primary and secondary variables; the variance of the secondary variable is set at 1.15. Another approach to integrating the secondary variable consists in modeling the local mean of the primary attribute by a linear regression of the secondary variable. This local mean is then input to a kriging with local varying mean to estimate the primary variable. Figure 7.8 shows the results of the cokriging and LVM approaches, all common parameters between the two algorithms being identical.

7.4 *BKRIG*: block kriging estimation

BKRIG is an algorithm for kriging estimation conditioned to both point and block support data (Goovaerts, 1997, p.152; Liu, 2007). The theory of kriging with linear averaged (block) data is recalled in Section 3.6.3. In this section we present codes that deal with some implementation issues specific to using block data.

Data error incorporation

In practice, observed data D (whether point data D_P or block average data D_B) are often associated with some noise or error, coming from measurement and subjective data interpretation among others. Since interpreted block average data tend to be more affected by error, only errors associated with block data are considered here. Thus:

$$D_P(\mathbf{u}_\alpha) = Z(\mathbf{u}_\alpha) \tag{7.3}$$

$$D_B(\mathbf{v}_\alpha) = B(\mathbf{v}_\alpha) + R(\mathbf{v}_\alpha) \tag{7.4}$$

where $Z(\mathbf{u}_\alpha)$ is the point data value at location $\mathbf{u}_\alpha$. $B(\mathbf{v}_\alpha)$ is the "true" block data value at block location $\mathbf{v}_\alpha$ and $R(\mathbf{v}_\alpha)$ is the error term associated with it.

The error $R(\mathbf{v}_\alpha)$ may depend on the signal $B(\mathbf{v}_\alpha)$, an heteroscedastic situation, and it may also correlate from one block to another (Koch and Link, 1970; Bourgault, 1994). Here the block errors are assumed homoscedastic and not cross-correlated:

$$R(\mathbf{v}_\alpha) \perp B(\mathbf{v}_\beta), \ \forall \mathbf{v}_\alpha, \mathbf{v}_\beta \tag{7.5}$$

$$R(\mathbf{v}_\alpha) \perp R(\mathbf{v}_\beta), \ \forall \mathbf{v}_\alpha, \mathbf{v}_\beta. \tag{7.6}$$

Also, assume the following properties of the block data errors (Journel and Huijbregts, 1978; Liu and Journel, 2005; Hansen *et al.*, 2006):

- zero mean: $E\{R(\mathbf{v}_\alpha)\} = 0, \ \forall \mathbf{v}_\alpha$,
- known variance: $\text{Var}\{R(\mathbf{v}_\alpha)\} = \sigma_R^2(\mathbf{v}_\alpha)$, which could be obtained from a prior calibration; note that this variance can vary from block to block,
- hence the error covariance is a known diagonal covariance matrix:

$$\mathbf{C}_R = [\text{Cov}\{R(\mathbf{v}_\alpha), R(\mathbf{u}_\beta)\}] = \begin{cases} [\sigma_R^2(\mathbf{v}_\alpha)] \text{ if } \mathbf{v}_\alpha = \mathbf{v}_\beta \\ [0] \text{ if } \mathbf{v}_\alpha \neq \mathbf{v}_\beta \end{cases} \tag{7.7}$$

If the point data are assumed error-free, the three sub-matrices $\mathbf{C}_{PP}$, $\bar{\mathbf{C}}_{PB}$ and $\bar{\mathbf{C}}_{PB}^t$ in Eq. (3.30) are unchanged. Only errors associated with block data are considered, the block-to-block covariance in the matrix $\mathbf{K}$ of system (3.30) is then written:

$$\bar{\mathbf{C}}_{B_\alpha B_\beta} = \left[\text{Cov}\{D_B(\mathbf{v}_\alpha), D_B(\mathbf{v}_\beta)\}\right]$$
$$= \left[\bar{C}_B(\mathbf{v}_\alpha, \mathbf{v}_\beta) + 2\bar{C}_{BR}(\mathbf{v}_\alpha, \mathbf{v}_\beta) + C_R(\mathbf{v}_\alpha, \mathbf{v}_\beta)\right]. \tag{7.8}$$

If the data errors are assumed independent of signal (Eq. (7.5)) and uncorrelated (Eq. (7.6)), and if the errors are assumed to have known variance $\sigma_R^2(\mathbf{v}_\alpha)$ (see Eq. (7.7)), expression (7.8) then becomes:

$$\bar{\mathbf{C}}_{B_\alpha B_\beta} = \begin{cases} \left[\bar{C}_B(0) + \sigma_R^2(\mathbf{v}_\alpha)\right] & \text{if } \mathbf{v}_\alpha = \mathbf{v}_\beta \\ \left[\bar{C}_B(\mathbf{v}_\alpha, \mathbf{v}_\beta)\right] & \text{if } \mathbf{v}_\alpha \neq \mathbf{v}_\beta. \end{cases} \tag{7.9}$$

As for the covariance vector $\mathbf{k}$ in Eq. (3.30), it is not affected by the error variance since the errors are assumed independent of the signals.

Therefore, the error impact can be incorporated into the kriging system by adding the error variances, $\sigma_R^2(\mathbf{v}_\alpha)$, as additional diagonal terms in the data-to-data covariance matrix on the left-hand side of the kriging system (3.30).

Point and block covariance calculation

Four types of covariances are needed in kriging (and simulation) algorithms involving block data (such as *BKRIG*, *BSSIM* in Section 8.1.7 and *BESIM* in Section 8.1.8): the point-to-point covariance $C_{PP'}$, the point-to-block covariance $\bar{C}_{PB}$, the block-to-point covariance $\bar{C}_{BP}$ and the block-to-block covariance $\bar{C}_{BB'}$, see kriging system Eq. (3.30). The point covariance $C_{PP'}$ is obtained through a precomputed covariance look-up table; the block average covariances ($\bar{C}_{PB}$, $\bar{C}_{BP}$ and $\bar{C}_{BB'}$) are computed through either a traditional integration method or an FFT-integration method.

Point covariance look-up table In algorithms such as *KRIGING* (Section 7.1), the covariance between any two points is calculated through the specified analytical variogram or covariance model. Instead of repeatedly calculating such covariance values, calculations can be done only once and the results stored in a point covariance look-up table. This point covariance look-up table is implemented as a covariance map (Goovaerts, 1997, p.99; Deutsch and Journel, 1998, p.36). For example, for a 2D field of size $M \times N$, all point covariances possibly used are contained in a covariance map of size $2M \times 2N$ with at its center the variance value $C(0)$. This covariance map is computed only once and stored. Look-up is performed whenever a covariance value is needed.

Two approaches for block covariance calculation A fast yet accurate block average covariance calculation is critical for any geostatistical algorithm involving block data. Two different approaches are proposed: the traditional integration method and an FFT-integration hybrid method.

In the traditional method, the block average covariance is calculated by averaging the point covariances. In its discrete form, this is written as (Journel and Huijbregts, 1978; Goovaerts, 1997, p.156):

$$\bar{C}_{PB} = \frac{1}{n} \sum_{i=1}^{n} C_{PP_i}$$

$$\bar{C}_{BB'} = \frac{1}{nn'} \sum_{i=1}^{n} \sum_{j=1}^{n'} C_{P_i P'_j} \qquad (7.10)$$

where $\bar{C}_{PB}$ is the covariance between point P and block B, $\bar{C}_{BB'}$ is the average covariance between block B and B'. P_i is one of the n point nodes discretizing block B and P'_j is one of the n' point nodes discretizing block B'. With a large number of block data and with dense block discretization, such direct calculation of block covariance would be very CPU expensive. Test results show that the block covariance calculation could account for more than 90% of the total simulation time. However, if there is a small number of blocks, this traditional integration approach is acceptable.

The FFT-integration hybrid method is a much more efficient algorithm. The basic idea of this algorithm is as follows. A symmetric circulant covariance matrix with circulant block sub-matrices is constructed from the covariance model and matrix. This circulant block covariance has many interesting features. Most notably, the whole covariance matrix can be fully retrieved by the single first row of block sub-matrices, entailing low memory cost; the multiplication with other matrix or vector can be achieved by a spectral convolution approach, entailing fast computations. Similarly, the block-to-point covariance map can be obtained by fast matrix multiplication via Fourier transform. Next, a classical averaging is applied to calculate the block-to-block covariances. See more details about this method in Dietrich and Newsam (1993); Nowak *et al.* (2003); Kyriakidis *et al.* (2005); Liu *et al.* (2006a). Availability of a fast FFT program is critical: the program **FFTW**, developed by Frigo and Johnson (2005), is used.

The basic implementation work-flow for computing the block-to-block covariance using this hybrid method is given in Algorithm 7.5.

Block data input

All block information is imported from a file. The file format is shown in Fig. 7.9. The first line is a description line giving any general information about block data. The second line is the number of blocks. Information related to each block follows. For each block section, the first line is the block name, followed by the block average value and the block error variance, and finally the coordinates of the points discretizing the block.

Algorithm 7.5 Block covariance calculation using FFT-integration
hybrid method

1: Generate the point covariance map $C_{P_\alpha P_\beta}$ with size double that of the field size
    in all x, y and z directions
2: Shift diagonally the quadrants of the extended covariance map
3: Extend the geometry of block #1 B_1 to the same size as the extended covariance
    map by zero-padding
4: Perform FFT on these two extended maps and multiply the two FFT results
5: Perform inverse FFT on the multiplication result to get the covariance map
    $\bar{C}_{B_1 P_\beta}$ between B_1 and any point P_β within the field
6: Average the previous block #1-to-point covariance values over the location of
    block #2 B_2. This gives the covariance $\bar{C}_{B_1 B_2}$ between B_1 and B_2

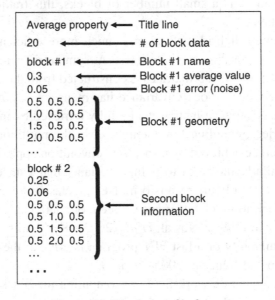

Figure 7.9 Block data file format

Block data reproduction

Per kriging theory, all point and block data are reproduced exactly if the data search
neighborhood includes them all. In practice, a larger search neighborhood and a
longer correlation range improve data reproduction.

Kriging type

Two types of kriging are accepted in *BKRIG*: simple kriging (SK) and ordinary
kriging (OK). The algorithm of *BKRIG* is given in Algorithm 7.6.

| **Algorithm 7.6** | Block kriging estimation |

1: Generate and store the point-to-point covariance look-up table. If the FFT-integration hybrid block covariance calculation method is used, compute and store the block-to-point covariance map(s)

2: **for** Each location **u do**

3:     Search the conditioning data consisting of the closest original point data and block data

4:     **if** the number of data (point or block) is large enough **then**

5:         Compute or retrieve the needed local block-to-block, block-to-point, point-to-block and point-to-point covariance

6:         Build and solve the mixed-scale kriging system

7:         Compute kriging mean and variance for location **u**

8:     **else**

9:         Set node as uninformed and issue warning message

10:     **end if**

11: **end for**

Parameters description

The *BKRIG* algorithm is activated from *Estimation | bkrig* in the algorithm panel. The main *BKRIG* interface contains three pages: "General", "Data" and "Variogram" (see Fig. 7.10). The text inside "[ ]" is the corresponding keyword in the *BKRIG* parameter file.

1. **Grid Name** [Grid_Name] Name of the estimation grid.
2. **Property Name Prefix** [Property_Name] Prefix for the estimation output.
3. **Kriging Type** [Kriging_Type] Select the type of Kriging system to be solved at each node: **Simple Kriging (SK)** or **Ordinary Kriging (OK)**.
4. **SK mean** [SK_mean] Mean of the attribute. Only required if **Kriging Type** [Kriging_Type] is set to **Simple Kriging (SK)**.
5. **Block Covariance Computation Approach** [Block_Cov_Approach] Select the method of computing block covariance: **FFT with Covariance-Table** or **Integration with Covariance-Table**.
6. **Check block data reproduction** [Check_Block_Reproduction] If checked, the estimated block average values are calculated and the relative errors compared to the input block data are computed for each realization. The results are shown on the **Commands Panel**, which is activated from *View → Commands Panel*.

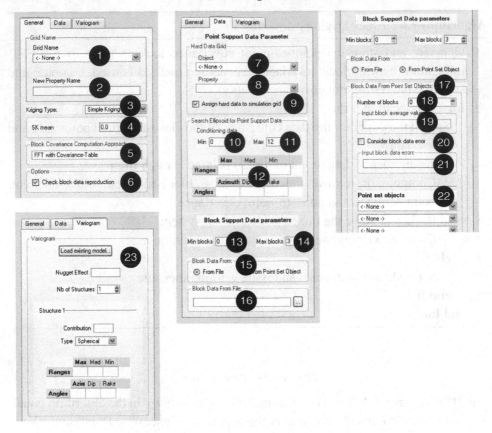

Figure 7.10 User interface for *BKRIG*

7. **Hard Data | Object** [Hard_Data.grid] Name of the grid containing the conditioning point data. If no point grid is selected, the estimation is performed conditioned only to block data.

8. **Hard Data | Property** [Hard_Data.property] Property for the point data. Only required if a grid has been selected in **Hard Data | Object** [Hard_Data.grid].

9. **Assign hard data to simulation grid** [Assign_Hard_Data] If selected, the hard data are relocated onto the estimation grid.

10. **Min conditioning point data** [Min_Conditioning_Data_Point] Minimum number of point data to be retained in the search neighborhood.

11. **Max conditioning point data** [Max_Conditioning_Data_Point] Maximum number of point data to be retained in the search neighborhood.

12. **Search Ellipsoid for Point Support Data** [Search_Ellipsoid_Point] Parametrization of the search ellipsoid for point support data, see Section 6.4.

13. **Min conditioning block data** [Min_Conditioning_Data_Block] Minimum number of block data to be retained in the search neighborhood.

14. **Max conditioning block data** [Max_Conditioning_Data_Block] Maximum number of block data to be retained in the search neighborhood.

15. **Block Data From** Select where the block data are to be found. There are two options: **From File** [Block_From_File] and **From Point Set Object** [Block_From_Pset].

16. **Block Data From File** [Block_Data_File] Only activated if **From File** [Block_From_File] is selected in **Block Data From**. The directory address of the block data file should be specified. The block data file format is shown in Fig. 7.9. If no block data file is entered, the estimation is performed using only point data.

17. **Block Data From Point Set Objects** Only activated if **From Point Set Object** [Block_From_Pset] is selected in **Block Data From**.

18. **Number of blocks** [Number_of_Blocks] Number of blocks entered from point-set objects.

19. **Input block average values** [Block_Average_Values] Enter the input block average value for each block. The sequence of block data should be the same as that of the corresponding input point-set objects. The number of block average values should be the same as the **Number of blocks** [Number_of_Blocks].

20. **Consider block data error** [Consider_Block_Error] If checked, the block errors are considered.

21. **Input block data errors** [Block_Error_Variance] Only activated if **Consider block data error** [Consider_Block_Error] is checked. Enter the block error variance for each block. The sequence of block data should be the same as that of the corresponding input point-set objects. The number of error variances should be the same as the **Number of blocks** [Number_of_Blocks].

22. **Point set objects** [Block_Grid_i] Enter the point-set block objects. This allows users to conveniently use the pre-loaded point-set objects as the conditioning blocks. No property is required to be associated with the input point-set grids. The maximum number of grids entered in this way is 50. They have to be loaded from a file if there are more than 50 blocks.

23. **Variogram parameters for simulation** [Variogram_Cov] Parametrization of the point-support variogram, see Section 6.5.

Examples

BKRIG is run for two 2D synthetic cases corresponding to tomography and downscaling applications. The reference model and input data for these two cases are given in Fig. 7.11.

The reference field for the tomography case is discretized into 40×40 grid cells, each cell of dimension 0.025×0.025. The reference background field has been

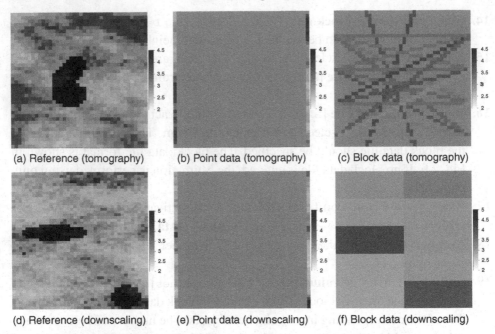

(a) Reference (tomography)   (b) Point data (tomography)   (c) Block data (tomography)

(d) Reference (downscaling)   (e) Point data (downscaling)   (f) Block data (downscaling)

Figure 7.11 The reference field, the point and block data for the tomography and the downscaling examples

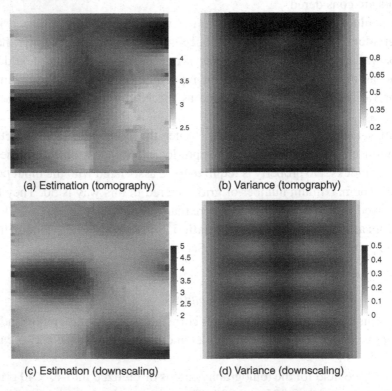

(a) Estimation (tomography)   (b) Variance (tomography)

(c) Estimation (downscaling)   (d) Variance (downscaling)

Figure 7.12 Kriging estimation results from *BKRIG*

generated using sequential Gaussian simulation (Section 3.8.1) with the normal
score variogram model:

$$\gamma(h_x, h_y) = 0.1 + 0.9 Sph \left(\sqrt{\left(\frac{h_x}{0.5}\right)^2 + \left(\frac{h_y}{0.25}\right)^2} \right). \tag{7.11}$$

A high value heterogeneous area is added in the center of the simulated field, see
Fig. 7.11a. The two columns of values at the left and right hand sides of the reference model are retained as conditioning point data, see Fig. 7.11b. The block data
are the 18 ray data, see their geometry in Fig. 7.11c. Each of the block datum value
is obtained by averaging point values over the ray path.

The reference field for the downscaling case study is discretized into 40×50
cells, each cell of dimension 0.025×0.02. The background property is generated
the same way as in the tomography case. Two high value heterogeneities are added
into that background field. Figure 7.11d gives the reference model. Again, the two
columns of values at the left and right hand sides of the reference model are retained
as conditioning point data, see Fig.7.11e. The block data are 10 coarse grid data
covering the entire field, see Fig.7.11f. Each of the block datum values is obtained
by averaging point values over the block.

Variogram model (Eq. (7.11)) is used for the *BKRIG* examples. The input SK
mean for the tomography case is 3.0 and that for the downscaling case is 2.7. The
minimum and maximum number of conditioning data are 0 and 12, respectively.

Figure 7.12a and Fig. 7.12c show the smooth kriging estimation maps using
both point and block data. The general patterns of the reference models are well
reflected, for example the locations of the high value heterogeneities. Figure 7.12b
gives the kriging variance for the tomography case. Low variances are found in
areas close to conditioning data. Figure 7.12d gives the kriging variance for the
downscaling case. In the middle area, there is less data constraint thus the variances
are high. The block data in both cases are well reproduced; the average absolute
errors are 1.8% and 0.5%, respectively.

8

Stochastic simulation algorithms

This chapter presents the SGeMS collection of stochastic simulation algorithms.

Section 8.1 presents the traditional variogram-based (two-point) algorithms, *SGSIM* (sequential Gaussian simulation), *SISIM* (sequential indicator simulation), *COSGSIM* (sequential Gaussian co-simulation), *COSISIM* (sequential indicator co-simulation), *DSSIM* (direct sequential simulation), *BSSIM* (block sequential simulation) and *BESIM* (block error simulation). *SGSIM*, *COSGSIM* and *DSSIM* are the choices for most continuous variables; *SISIM* and *COSISIM* are designed for categorical variables; *BSSIM* and *BESIM* are developed for simulation conditioning to block averaged data and point data.

Section 8.2 gives detailed descriptions of two multiple-point statistics (MPS) algorithms: *SNESIM* (single normal equation simulation) and *FILTERSIM* (filter-based simulation). *SNESIM* works best for categorical variables such as facies distributions. Two algorithms cover the *FILTERSIM* paradigm; *FILTERSIM_CONT* for continuous attributes and *FILTERSIM_CATE* for categorical attributes. The *FILTERSIM* framework is better suited for continuous variables but also gives good results with categorical variables.

Each simulation algorithm presented in this chapter is demonstrated with an example.

8.1 Variogram-based simulations

This section covers the variogram-based sequential simulation algorithms implemented in SGeMS. Simulated realizations from any of these algorithms draw their spatial patterns from input variogram models. Variogram-based algorithms should preferably be used to simulate reasonably amorphous (high entropy) distributions. Cases where variograms fail to model specific spatial patterns call for using the multiple-point geostatistics algorithms described in the next section.

Variogram-based sequential simulations have been the most popular stochastic simulation algorithms mostly due to their robustness and ease of conditioning, both to hard and soft data. Moreover, they do not require rasterized (regular or Cartesian) grids; they allow simulation on irregular grids such as point-sets.

SGeMS keeps this flexibility; all variogram-based simulation algorithms described in this section work both on point-sets and on Cartesian grids. The conditioning data may or may not be on the simulation grid. However, working on point-set induces a performance penalty as the search for neighboring data is significantly more costly than on a regular (Cartesian) grid. When the simulation grid is Cartesian, all algorithms have the option to relocate the conditioning data to the nearest grid node for increased execution speed. The re-allocation strategy is to move each datum to the closest grid node. In case two data share the same closest grid node, the further one is ignored.

This section presents first the simulation algorithms requiring a Gaussian assumption: LU simulation *LUSIM*, sequential Gaussian simulation *SGSIM* and sequential Gaussian co-simulation *COSGSIM* for integration of secondary information through a coregionalization model. Next, direct sequential simulation *DSSIM* is presented; *DSSIM* does not require any Gaussian assumption, thus works on the original variable without any preliminary normal score transform. Next, the two indicator simulation algorithms *SISIM* and *COSISIM* are documented. For continuous variables, indicator algorithms rely on the discretization of the cumulative distribution function by a set of threshold values. At any location the probability of not exceeding each threshold is estimated by indicator kriging, these probabilities are then combined to construct the local conditional cumulative distribution function (ccdf) from which a simulated value of the continuous variable is drawn. Last, algorithms *BESIM* and *BSSIM*, which can be used to account for linear block averages, are presented.

8.1.1 LUSIM: *LU simulation*

The LU simulation algorithm is a multi-Gaussian algorithm appropriate for small data sets (Deutsch and Journel, 1998, p.146). *LUSIM* is an exact algorithm to simulate a Gaussian random field; it performs a Cholesky decomposition of the covariance matrix (Davis, 1987). The advantage is that this LU decomposition only needs to be performed once, each additional realization is generated at the cost of a mere matrix multiplication. The major drawback is that the LU decomposition, even if done only once, is a very costly operation, hence should be restricted to small fields and data sets (Dimitrakopoulos and Luo, 2004).

The LU simulation is the equivalent of the sequential Gaussian simulation algorithm with simple kriging and with an infinite data search neighborhood

(Alabert, 1987; Dimitrakopoulos and Luo, 2004); all original conditioning data and previously simulated values are used in the kriging system at every node along the simulation path. The LU simulation algorithm is detailed in Algorithm 8.1.

Algorithm 8.1 LU simulation

1: Transform the data into normal score space. $Z(\mathbf{u}) \mapsto Y(\mathbf{u})$
2: Build the Covariance matrix
3: Perform LU decomposition of the covariance matrix using the Cholesky decomposition
4: Multiply a vector of independent Gaussian random deviates with the result of step 3
5: Back transform the Gaussian simulated field into the data space. $Y(\mathbf{u}) \mapsto Z(\mathbf{u})$
6: Repeat from step 4 for another realization

Parameters description

The *LUSIM* algorithm is activated from *Simulation* → *lusim* in the Algorithm Panel. The *LUSIM* interface contains three pages: "General", "Data" and "Variogram" (see Fig. 8.1). The text inside "[ ]" is the corresponding keyword in the *LUSIM* parameter file.

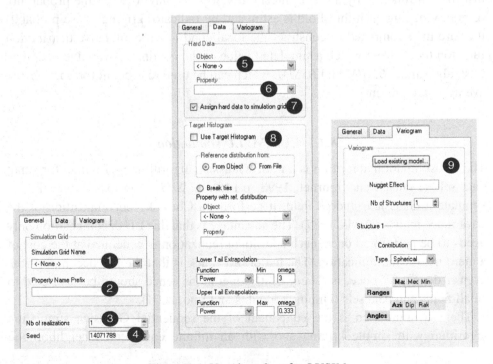

Figure 8.1 User interface for *LUSIM*

1. **Simulation Grid Name** [Grid_Name]   Name of the simulation grid.
2. **Property Name Prefix** [Property_Name]   Prefix for the simulation output. The suffix __real# is added for each realization.
3. **# of realizations** [Nb_Realizations]   Number of simulations to generate.
4. **Seed** [Seed]   Seed for the random number generator (preferably a large odd integer).
5. **Hard Data | Object** [Hard_Data.grid]   Name of the grid containing the conditioning data. If no grid is selected, the realizations are unconditional.
6. **Hard Data | Property** [Hard_Data.property]   Property for the conditioning data. Only required if a grid has been selected in **Hard Data | Object** [Hard_Data.grid].
7. **Assign hard data to simulation grid** [Assign_Hard_Data]   If selected, the hard data are relocated to the simulation grid. The program does not proceed if the assignment fails. This option significantly increases execution speed.
8. **Target Histogram**   If used, the data are normal score transformed prior to simulation and the simulated field is transformed back to the original space. **Use Target Histogram** [Use_Target_Histogram] flag to use the normal score transform. The target histogram is parametrized by [nonParamCdf], see Section 6.8.
9. **Variogram** [Variogram]   Parametrization of the normal score variogram, see Section 6.5.

8.1.2 SGSIM: *sequential Gaussian simulation*

The *SGSIM* algorithm uses the sequential simulation formalism to simulate a Gaussian random function. Let $Y(\mathbf{u})$ be a multivariate Gaussian random function with zero mean, unit variance, and a given variogram model $\gamma(\mathbf{h})$. Realizations of $Y(\mathbf{u})$ can be generated by Algorithm 8.2.

Algorithm 8.2   Sequential Gaussian simulation

1: Define a random path visiting each node of the grid
2: **for** Each node $\mathbf{u}$ along the path **do**
3:    Get the conditioning data consisting of neighboring original hard data (n) and previously simulated values
4:    Estimate the local conditional cdf as a Gaussian distribution with mean given by kriging and variance by the kriging variance
5:    Draw a value from that Gaussian ccdf and add the simulated value to the data set
6: **end for**
7: Repeat for another realization

A non-Gaussian random function $Z(\mathbf{u})$ must first be transformed into a Gaussian random function $Y(\mathbf{u})$; $Z(\mathbf{u}) \mapsto Y(\mathbf{u})$. If no analytical model is available, a normal score transform should be applied. The simulated Gaussian values should then be back transformed. Algorithm 8.2 then becomes Algorithm 8.3.

Algorithm 8.3   Sequential Gaussian simulation with normal score transform

---

1: Transform the data into normal score space. $Z(\mathbf{u}) \mapsto Y(\mathbf{u})$
2: Perform Algorithm 8.2
3: Back transform the Gaussian simulated field into the data space. $Y(\mathbf{u}) \mapsto Z(\mathbf{u})$

The algorithm calls for the variogram of the normal score *not* of the original data. Only that normal score variogram is guaranteed to be reproduced within ergodic fluctuations, not the original Z-value variogram. However, in most cases the back transform does not adversely affect the reproduction of the Z-value variogram. If required, the SGeMS implementation of *SGSIM* can automatically perform the normal score transform of the original hard data and back-transform the simulated realizations. The user must still independently perform the normal score transform of the hard data with program *TRANS*, see Algorithm 9.1, in order to compute and model the normal score variogram.

At each simulation grid node, to determine the Gaussian ccdf, the algorithm can use either simple kriging, ordinary kriging, kriging with local mean, or kriging with a trend. Theory guarantees variogram reproduction only when simple kriging is used.

SGSIM *with local varying mean*

In many applications, a local varying mean $z_m(\mathbf{u}) = E\{Z(\mathbf{u})\}$ may be available. It is given in Z-unit, thus must be converted into the Gaussian space, such that $y_m(\mathbf{u}) = E\{Y(\mathbf{u})\}$. Transforming $z_m(\mathbf{u})$ into $y_m(\mathbf{u})$ using the Z-marginal cdf F_Z and the rank preserving transform

$$y_m(\mathbf{u}) = G^{-1}\left(F_Z\Big(z_m(\mathbf{u})\Big) \right)$$

would not, in all rigor, ensure that $y_m(\mathbf{u}) = E\{Y(\mathbf{u})\}$. Here $G(\cdot)$ is the standard normal cdf and $F_Z(\cdot)$ is the z-target histogram. A better alternative is to infer the normal score varying mean $y_m(\mathbf{u})$ by some direct calibration of the secondary information to the normal score transform y of the primary attribute z.

A note on Gaussian spatial law

Gaussian random functions have very specific and consequential spatial structures and distribution law. For example, median values are maximally correlated in space, while extreme values are increasingly less correlated. That property is known as the destructuration effect (Goovaerts, 1997, p.272). If the phenomenon under study is known to have well correlated extreme values or high values correlated differently from low values, a Gaussian-related simulation algorithm is not appropriate.

Also, for a given covariance/variogram model the Gaussian random function model is that which maximizes spatial entropy, that is spatial disorder. Modeling organized systems with low entropy requires much more than a variogram, hence their simulation would call for multiple-point statistics beyond the 2-point statistics of a variogram; see Section 3.9.

Parameters description

The *SGSIM* algorithm is activated from *Simulation → sgsim* in the algorithm panel. The main *SGSIM* interface contains three pages: "General", "Data" and "Variogram" (see Fig. 8.2). The text inside "[ ]" is the corresponding keyword in the *SGSIM* parameter file.

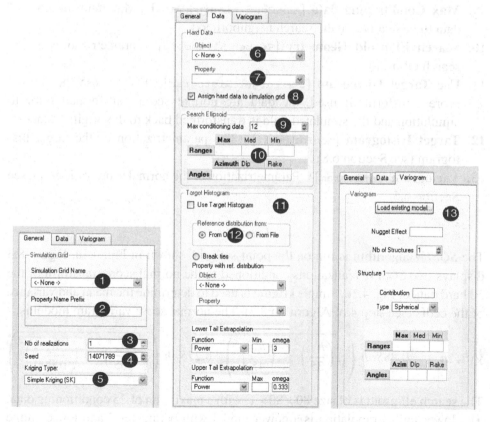

Figure 8.2 User interface for *SGSIM*

1. **Simulation Grid Name** [Grid_Name]   Name of the simulation grid.
2. **Property Name Prefix** [Property_Name]   Prefix for the simulation output. The suffix __real# is added for each realization.
3. **# of realizations** [Nb_Realizations]   Number of simulations to generate.
4. **Seed** [Seed]   Seed for the random number generator (preferably a large odd integer).
5. **Kriging Type** [Kriging_Type]   Select the type of kriging system to be solved at each node along the random path. The simple kriging (SK) mean is set to zero, as befits a stationary standard Gaussian model.
6. **Hard Data | Object** [Hard_Data.grid]   Name of the grid containing the conditioning data. If no grid is selected, the realizations are unconditional.
7. **Hard Data | Property** [Hard_Data.property]   Property for the conditioning data. Only required if a grid has been selected in **Hard Data | Object** [Hard_Data.grid].
8. **Assign hard data to simulation grid** [Assign_Hard_Data]   If selected, the hard data are relocated onto the simulation grid. The program does not proceed if the relocation fails. This option significantly increases execution speed.
9. **Max Conditioning data** [Max_Conditioning_Data]   Maximum number of data to be retained in the search neighborhood.
10. **Search Ellipsoid Geometry** [Search_Ellipsoid]   Parametrization of the search ellipsoid.
11. **Use Target Histogram** [Use_Target_Histogram]   Flag to use the normal score transform. If used, the data are normal score transformed prior to simulation and the simulated field is transformed back to the original space.
12. **Target Histogram** [nonParamCdf]   The parametrization of the target histogram (see Section 6.8).
13. **Variogram** [Variogram]   Parametrization of the normal score variogram, see Section 6.5.

Example

The *SGSIM* algorithm is run on the point-set grid, shown in Fig. 4.1a. Figure 8.3 displays two *SGSIM* realizations conditioned to the normal score transform of the 50 hard data of Fig. 4.1e. Simple kriging is used to determine the mean and variance of the ccdfs (see step 4 of Algorithm 8.2). The normal score variogram model is:

$$\gamma(h_x, h_y) = 0.88 Sph\left(\sqrt{\left(\frac{h_x}{35}\right)^2 + \left(\frac{h_y}{45}\right)^2}\right) + 0.27 Sph\left(\sqrt{\left(\frac{h_x}{65}\right)^2 + \left(\frac{h_y}{10000}\right)^2}\right).$$

The search ellipsoid is of size $80 \times 80 \times 1$, with a maximum of 25 conditioning data. The lower tail extrapolation is a power model with parameter 3 and low extreme

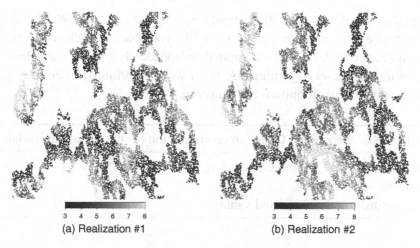

(a) Realization #1   (b) Realization #2

Figure 8.3 Two *SGSIM* realizations

value 3.4; the upper tail extrapolation is a power model with parameter 0.333 and a maximum value of 8.4.

8.1.3 COSGSIM: *sequential Gaussian co-simulation*

COSGSIM allows to simulate a Gaussian variable while accounting for secondary information. Let $Y_1(\mathbf{u})$ and $Y_2(\mathbf{u})$ be two correlated multi-Gaussian random variables. $Y_1(\mathbf{u})$ is called the primary variable, and $Y_2(\mathbf{u})$ the secondary variable. The *COSGSIM* simulation of the primary variable Y_1 conditioned to both primary and secondary data is described in Algorithm 8.4.

Algorithm 8.4   Sequential Gaussian co-simulation

1: Define a path visiting each node $\mathbf{u}$ of the grid
2: **for** Each node $\mathbf{u}$ **do**
3:   Get the conditioning data consisting of neighboring original data and previously simulated values and the secondary data
4:   Get the local Gaussian ccdf for the primary attribute, with mean equal to the cokriging estimate and variance equal to the cokriging variance
5:   Draw a value from that Gaussian ccdf and add it to the data set
6: **end for**

If the primary and secondary variables are not Gaussian, make sure that each of the transformed variables Y_1 and Y_2 is at least univariate Gaussian. If they are not, another simulation algorithm should be considered, *COSISIM* for example (see Section 8.1.6). If no analytical model is available for such transformation, a normal score transform may be applied independently to both variables. Algorithm 8.4 then becomes Algorithm 8.5. The *TRANS* algorithm only ensures that the respective marginal distributions are Gaussian.

Algorithm 8.5 Sequential Gaussian co-simulation for non-Gaussian variable

---

1: Transform Z_1 and Z_2 into Gaussian variables Y_1 and Y_2
2: Perform Algorithm 8.4
3: Back-transform the simulated values

Parameters description

The *COSGSIM* algorithm is activated from *Simulation* → *cosgsim* in the algorithm panel. The *COSGSIM* interface contains five pages: "General", "Primary Data", "Secondary Data", "Primary Variogram" and "Secondary Variogram" (see Fig. 8.4). The text inside "[ ]" is the corresponding keyword in the *COSGSIM* parameter file.

1. **Simulation Grid Name** [Grid_Name]   Name of the simulation grid.
2. **Property Name Prefix** [Property_Name]   Prefix for the simulation output. The suffix __real# is added for each realization.
3. **# of realizations** [Nb_Realizations]   Number of simulations to generate.
4. **Seed** [Seed]   Seed for the random number generator (preferably a large odd integer).
5. **Kriging Type** [Kriging_Type]   Select the type of kriging system to be solved at each node along the random path.
6. **Cokriging Option** [Cokriging_Type]   Select the type of coregionalization model: LMC, MM1 or MM2.
7. **Primary Hard Data Grid** [Primary_Harddata_Grid]   Selection of the grid for the primary variable. If no grid is selected, the realizations are unconditional.
8. **Primary Property** [Primary_Variable]   Selection of the hard data property for the primary variable.
9. **Assign hard data to simulation grid** [Assign_Hard_Data]   If selected, the hard data are relocated onto the simulation grid. The program does not proceed if the relocation fails. This option significantly increases execution speed.
10. **Primary Max Conditioning data** [Max_Conditioning_Data_1]   Maximum number of primary data to be retained in the search neighborhood.

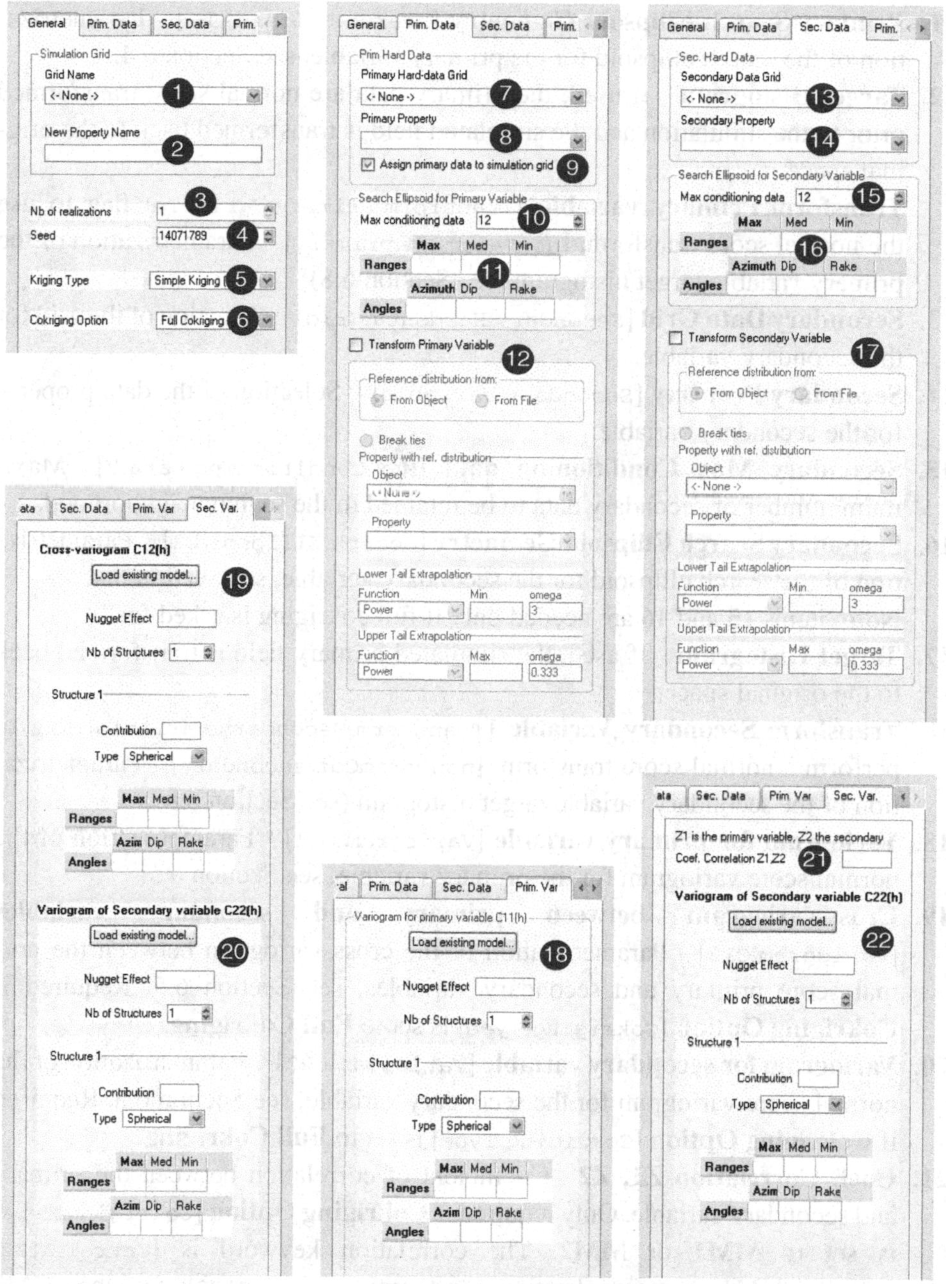

Figure 8.4 User interface for *COSGSIM*

11. **Primary Search Ellipsoid Geometry** [Search_Ellipsoid_1] Parametrization of the search ellipsoid for the primary variable, see Section 6.4.

12. **Target Histogram** If used, the primary data are normal score transformed prior to the simulation and the simulated field is transformed back to the original space.
    Transform_Primary_Variable [Transform_Primary_Variable] flag to use the normal score transform. [nonParamCdf_primary] Parametrization of the primary variable target histogram (see Section 6.8).

13. **Secondary Data Grid** [Secondary_Harddata_Grid] Selection of the grid for the secondary variable.

14. **Secondary Property** [Secondary_Variable] Selection of the data property for the secondary variable.

15. **Secondary Max Conditioning data** [Max_Conditioning_Data_2] Maximum number of secondary data to be retained in the search neighborhood.

16. **Secondary Search Ellipsoid Geometry** [Search_Ellipsoid_2] Parametrization of the search ellipsoid for the secondary variable, see Section 6.4.
    Note: items **15** and **16** are needed only if full cokriging is asked for.

17. **Target Histogram** If used, the simulated primary field is transformed back to the original space.
    Transform_Secondary_Variable [Transform_Secondary_Variable] flag to perform a normal score transform. [nonParamCdf_secondary] Parametrization of the secondary variable target histogram (see Section 6.8).

18. **Variogram for primary variable** [Variogram_C11] Parametrization of the normal score variogram for the primary variable, see Section 6.5.

19. **Cross-variogram between primary and secondary variables** [Variogram_C12] Parametrization of the cross-variogram between the normal score primary and secondary variables, see Section 6.5. Required if **Cokriging Option** [Cokriging_Type] is set to **Full Cokriging**.

20. **Variogram for secondary variable** [Variogram_C22] Parametrization of the normal score variogram for the secondary variable, see Section 6.5. Required if **Cokriging Option** [Cokriging_Type] is set to **Full Cokriging**.

21. **Coef. Correlation Z1, Z2** Coefficient of correlation between the primary and secondary variable. Only required if **Cokriging Option** [Cokriging_Type] is set to **MM1** or **MM2**. The correlation keyword is [Correl_Z1Z2] for the MM1 coregionalization, and [MM2_Correl_Z1Z2] for the MM2 coregionalization.

22. **Variogram for secondary variable** [MM2_Variogram_C22] Parametrization of the normal score variogram for the secondary variable, see Section 6.5. Required if **Cokriging Option** [Cokriging_Type] is set to **MM2**.

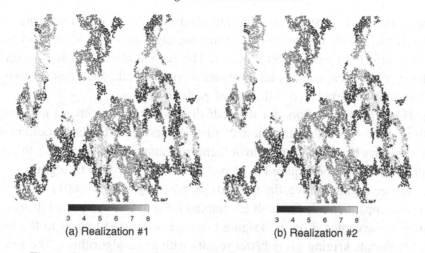

(a) Realization #1   (b) Realization #2

Figure 8.5 Two *COSGSIM* realizations with MM1 coregionalization

Example

The *COSGSIM* algorithm is run on the point set grid shown in Fig. 4.1a. Two conditional *COSGSIM* realizations with an MM1-type of co-regionalization and simple cokriging are shown in Fig. 8.5. Both the primary hard conditioning data (Fig. 4.1e) and secondary information (Fig. 4.1d) were normal score transformed. The normal score variogram model for the primary variable is:

$$\gamma(h_x, h_y) = 0.88 Sph\left(\sqrt{\left(\frac{h_x}{35}\right)^2 + \left(\frac{h_y}{45}\right)^2}\right) + 0.27 Sph\left(\sqrt{\left(\frac{h_x}{65}\right)^2 + \left(\frac{h_y}{10\,000}\right)^2}\right),$$

and the correlation coefficient between the primary and secondary variables is 0.7. The search ellipsoid is of size $80 \times 80 \times 1$, with a maximum of 25 primary conditioning data. The lower tail extrapolation is a power model with parameter 3 and minimum value 3.4; the upper tail extrapolation is a power model with parameter 0.333 and maximum value 8.4.

8.1.4 DSSIM: *direct sequential simulation*

The direct sequential simulation algorithm *DSSIM* performs simulation of continuous attributes without prior indicator coding or Gaussian transform. As seen in Section 3.8.2, the only condition for the model variogram to be reproduced (within fluctuations) is that the ccdf has for mean and variance the simple kriging estimate and variance. The shape of the ccdf does not matter; it may not even be the same for each simulated node. The drawback is that there is no guarantee that the marginal distribution is reproduced (Journel, 1994).

One solution is to post-process the simulated realizations with a rank-preserving transform to identify the target histogram; see algorithm *TRANS* in Section 9.1. This may affect variogram reproduction. The second alternative is to determine the shape of the local ccdf, at all locations along the path, such that the marginal distribution is approximated at the end of each realization.

DSSIM offers two options for the ccdf distribution type, either a uniform distribution or a lognormal distribution. Neither of these distributions would produce realizations that have either uniform or lognormal marginal distributions, thus some post processing may be required to identify the target marginal histogram.

For the second alternative, the methods proposed by Soares (2001) and Oz *et al.* (2003) are implemented. The ccdf is sampled from the data marginal distribution, modified to be centered on the kriging estimate with spread equal to the kriging variance. Simple kriging gives better results with these algorithms. The resulting shape of each local ccdf thus differs from location to location. The method gives reasonable reproduction of a target symmetric (even multi-modal) distribution, but poorer results for highly skewed distributions. In this latter case the first option using a log-normal ccdf type followed by a final post-processing using *TRANS* may give better results. The general *DSSIM* algorithm is given in Algorithm 8.6.

Algorithm 8.6   Direct sequential simulation

1: Define a random path visiting each node **u** of the grid
2: **for** Each location **u** along the path **do**
3:     Get the conditioning data consisting of both neighboring original data and previously simulated values
4:     Define the local ccdf with its mean and variance given by the kriging estimate and variance
5:     Draw a value from that ccdf and add the simulated value to the data set
6: **end for**

Parameters description

The *DSSIM* algorithm is activated from *Simulation* → *dssim* in Algorithm Panel. The *DSSIM* interface contains three pages: "General", "Data" and "Variogram" (see Fig. 8.6). The text inside "[ ]" is the corresponding keyword in the *DSSIM* parameter file.

1. **Simulation Grid Name** [Grid_Name]   Name of the simulation grid.
2. **Property Name Prefix** [Property_Name]   Prefix for the simulation output. The suffix _real# is added for each realization.
3. **# of realizations** [Nb_Realizations]   Number of simulations to generate.

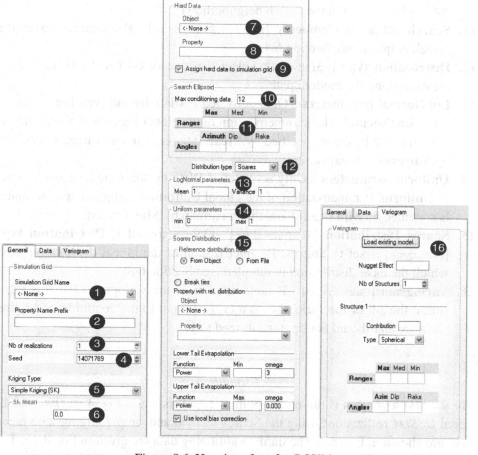

Figure 8.6 User interface for *DSSIM*

4. **Seed** [Seed]   Seed for the random number generator (preferably a large odd integer).

5. **Kriging Type** [Kriging_Type]   Select the form of kriging system to be solved at each node along the random path.

6. **SK_mean** [SK_mean]   Mean of the attribute. Only required if **Kriging Type** [Kriging_Type] is set to Simple Kriging (SK).

7. **Hard Data | Object** [Hard_Data.grid]   Name of the grid containing the conditioning data. If no grid is selected, the realizations are unconditional.

8. **Hard Data | Property** [Hard_Data.property]   Property for the conditioning data. Only required if a grid has been selected in **Hard Data | Object** [Hard_Data.grid].

9. **Assign hard data to simulation grid** [Assign_Hard_Data]   If selected, the hard data are copied on the simulation grid. The program does not proceed if the relocation fails. This option improves execution speed.

10. **Max Conditioning data** [Max_Conditioning_Data]   Maximum number of data to be retained in the search neighborhood.
11. **Search Ellipsoid Geometry** [Search_Ellipsoid]   Parametrization of the search ellipsoid, see Section 6.4.
12. **Distribution type** [cdf_type]   Select the type of ccdf to be build at each location along the random path.
13. **LogNormal parameters**   Only activated if **Distribution type** [cdf_type] is set to **LogNormal**. The parametrization of the global lognormal distribution is done through its mean specified by **Mean** [LN_mean] and its variance specified by **Variance** [LN_variance].
14. **Uniform parameters**   Only activated if **Distribution type** [cdf_type] is set to **Uniform**. Parametrization of the global Uniform distribution, the minimum is specified by **Min** [U_min] and the maximum by **Max** [U_max].
15. **Soares Distribution** [nonParamCdf]   Only activated if **Distribution type** [cdf_type] is set to **Soares**. Parametrization of the global distribution from which the local distribution is sampled (see Section 6.8).
16. **Variogram** [Variogram]   Parametrization of the variogram. For this algorithm, the sill of the variogram is a critical input to the conditioning kriging variance and should not be standardized to 1.

Example

The *DSSIM* algorithm is run on the point-set grid shown in Fig. 4.1a. Two conditional *DSSIM* realizations using the Soares method and simple kriging with mean 5.45 are shown in Fig. 8.7. The hard conditioning data are given in Fig. 4.1e. The variogram model (in the original data space) for the primary variable is:

$$\gamma(h_x, h_y) = 1.2 Sph\left(\sqrt{\left(\frac{h_x}{35}\right)^2 + \left(\frac{h_y}{45}\right)^2}\right) + 0.2 Sph\left(\sqrt{\left(\frac{h_x}{35}\right)^2 + \left(\frac{h_y}{100000}\right)^2}\right).$$

The search ellipsoid is of size $80 \times 80 \times 1$, with a maximum of 25 conditioning data. The lower tail extrapolation is a power model with parameter 3 and a minimum value 3.4; the upper tail extrapolation is a power model with parameter 0.333 and a maximum value 8.4.

The histogram plots of both the reference distribution and the *DSSIM* realization #1 are given in Fig. 8.8; for this example it appears that the *DSSIM* algorithm with Soares method does reproduce the target histogram reasonably well.

Figure 8.9 gives the two *DSSIM* realizations using Soares method and kriging with the local varying mean shown in Fig. 4.1b.

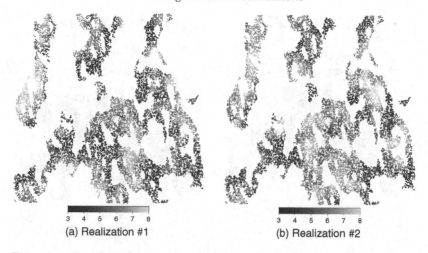

(a) Realization #1   (b) Realization #2

Figure 8.7 Two *DSSIM* realizations with Soares method and simple kriging

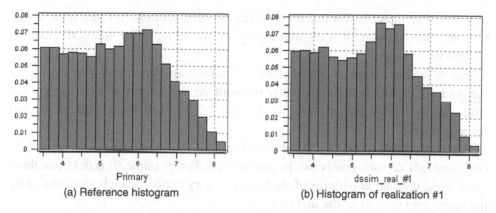

(a) Reference histogram   (b) Histogram of realization #1

Figure 8.8 Histograms of the reference and *DSSIM* realization #1

8.1.5 SISIM: *sequential indicator simulation*

Sequential indicator simulation *SISIM* combines the indicator formalism with the sequential paradigm to simulate non-parametric continuous or categorical distributions. In the continuous case, at every location along the simulation path, non-parametric ccdf values, one per threshold value, are estimated using indicator kriging from neighboring indicator-transformed data. For the categorical case, the probability for each category to occur is estimated by indicator kriging.

Sequential indicator simulation (Algorithm 8.7) does not require any Gaussian assumption. Instead, in the continuous case, the ccdf is built by estimating a sequence of probabilities of not exceeding a finite set of threshold values. The more thresholds retained, the more detailed the conditional cumulative distribution

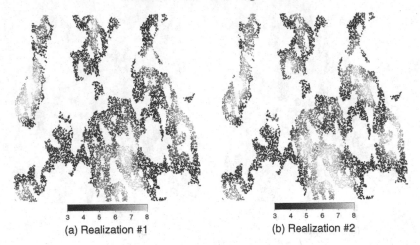

(a) Realization #1   (b) Realization #2

Figure 8.9 Two *DSSIM* realizations with Soares method and local varying mean

function. The indicator formalism removes the need for normal score transformation, but does require an additional variogram modeling effort unless the median IK option is chosen.

The SGeMS version of *SISIM* does not require any prior indicator coding of the data, all coding is done internally for both the continuous and the categorical cases. Interval or incomplete data, see Section 7.2, may also be entered but need to be pre-coded.

When modeling a set of indicator variograms, the user is warned that not all combinations of variograms can be reproduced. For example, if a field has three categories, the spatial patterns of the third category are completely determined by the variogram models of the first two.

Continuous variables

Algorithm *SISIM* relies on indicator kriging to infer the local ccdf values. The indicator variable corresponding to the z-continuous variable is defined as:

$$i(\mathbf{u}, z_k) = \begin{cases} 1 & \text{if } z(\mathbf{u}) \leq z_k \\ 0 & \text{otherwise} \end{cases}, \quad k = 1, \dots, K.$$

Two types of regionalizations can be used for indicator simulation. The full IK option requires a variogram model for each threshold z_k. The median IK (Goovaerts, 1997, p.304) option requires only the variogram model for the median threshold, all other indicator variograms are assumed proportional to that single model.

While both *SGSIM* and *SISIM* with median IK require a single variogram, they produce different outputs and spatial patterns. The major difference is that the extreme values of a *SISIM* realization with median IK are more spatially correlated than in a *SGSIM* realization.

SISIM can handle interval data the same way that *INDICATOR KRIGING* does, see Section 7.2.

For continuous attributes *SISIM* implements Algorithm 8.7.

Algorithm 8.7   *SISIM* with continuous variables

1: Choose a discretization of the range of $Z(\mathbf{u})$: $z_1, \ldots, z_K$
2: Define a path visiting all locations to be simulated
3: **for** Each location $\mathbf{u}$ along the path **do**
4:   Retrieve the neighboring conditioning data: $z(\mathbf{u}_\alpha), \alpha = 1, \ldots, n(\mathbf{u})$
5:   Turn each datum $z(\mathbf{u}_\alpha)$ into a vector of indicator values:
   $$\mathbf{i}(\mathbf{u}_\alpha) = \left[i(\mathbf{u}_\alpha, z_1), \ldots, i(\mathbf{u}_\alpha, z_K) \right]$$
6:   Estimate the indicator random variable $I(\mathbf{u}, z_k)$ for each of the K thresholds by solving a kriging system
7:   After correction of order relation deviations, the estimated values $i^*(\mathbf{u}, z_k) = Prob^*(Z(\mathbf{u}) \leq z_k | (n(\mathbf{u})))$ define an estimate of the ccdf $F_{Z(\mathbf{u})}$ of the variable $Z(\mathbf{u})$
8:   Draw a value from that ccdf and assign it as a datum at location $\mathbf{u}$
9: **end for**
9: Repeat the previous steps to generate another simulated realization

---

Sampling the estimated distribution function

At each location to be simulated, the ccdf $F(\mathbf{u}; z_k | (n))$ is estimated at the K thresholds $z_1, \ldots, z_K$. However, sampling from such ccdf, as described in step 7 of Algorithm 8.7, requires knowledge of the ccdf $F(\mathbf{u}; z | (n))$ for all values z. In *SISIM* the ccdf is interpolated as follows:

$$F^*(\mathbf{u}; z) = \begin{cases} \phi_{\text{lti}}(z), & \text{if } z \leq z_1 \\ F^*(\mathbf{u}; z_k) + \frac{z - z_k}{z_{k+1} - z_k}\left(F^*(\mathbf{u}; z_{k+1}) - F^*(\mathbf{u}; z_k)\right), & \text{if } z_k \leq z < z_{k+1} \\ 1 - \phi_{\text{uti}}(z), & \text{if } z_{k+1} \leq z \end{cases}$$

(8.1)

where $F^*(\mathbf{u}; z_k) = i^*(\mathbf{u}, z_k)$ is obtained by indicator kriging and $\phi_{\text{lti}}(z)$ and $\phi_{\text{uti}}(z)$ are the lower and upper tail extrapolation chosen by the user and described in Section 6.8. Values between thresholds z_i and z_{i+1} are interpolated linearly, hence are drawn from a uniform distribution in the interval $[z_i, z_{i+1}]$.

Categorical variables

If $Z(\mathbf{u})$ is a categorical variable that only takes the K integer values $\{0, \ldots, K-1\}$, Algorithm 8.7 is modified as described in Algorithm 8.8.

The categorical indicator variables are defined as:

$$i(\mathbf{u}, k) = \begin{cases} 1 & \text{if } Z(\mathbf{u}) = k \\ 0 & \text{otherwise.} \end{cases}$$

For the categorical case, the median IK option implies that all categories share the same variogram up to a proportionality factor.

Algorithm 8.8   *SISIM* with categorical variable

1: Define a path visiting all locations to be simulated
2: **for** Each location $\mathbf{u}$ along the path **do**
3:     Retrieve the neighboring categorical conditioning data:
       $z(\mathbf{u}_\alpha), \alpha = 1, \ldots, n(\mathbf{u})$
4:     Turn each datum $z(\mathbf{u}_\alpha)$ into a vector of indicator data values:
       $\mathbf{i}(\mathbf{u}_\alpha) = \left[i(\mathbf{u}_\alpha, z_1), \ldots, i(\mathbf{u}_\alpha, z_K) \right]$
5:     Estimate the indicator random variable $I(\mathbf{u}, k)$ for each of the K categories
       by solving a simple kriging system
6:     After correction of order relation deviations, the estimated values $i^*(\mathbf{u}, k) = Prob^*(Z(\mathbf{u}) = k|(n))$ define an estimate of the discrete conditional probability density function (cpdf) of the categorical variable $Z(\mathbf{u})$
7:     Draw a simulated category from that cpdf and assign it as a datum at location $\mathbf{u}$
8: **end for**
9: Repeat the previous steps to generate another realization

Parameters description

The *SISIM* algorithm is activated from *Simulation* → *sisim* in the Algorithm Panel. The main *SISIM* interface contains three pages: "General", "Data" and "Variogram" (see Fig. 8.10). The text inside "[ ]" is the corresponding keyword in the *SISIM* parameter file.

1. **Simulation Grid Name** [Grid_Name]   Name of the simulation grid.
2. **Property Name Prefix** [Property_Name]   Prefix for the simulation output. The suffix __real# is added for each realization.
3. **# of realizations** [Nb_Realizations]   Number of simulations to generate.
4. **Seed** [Seed]   Seed for the random number generator (preferably a large odd integer).

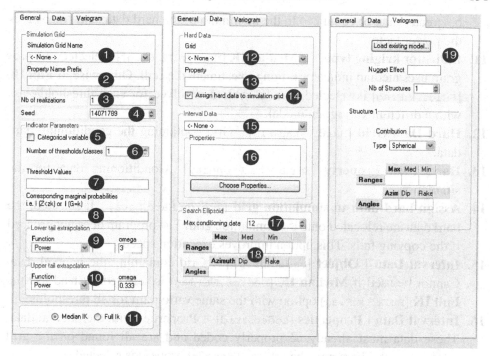

Figure 8.10 User interface for *SISIM*

5. **Categorical variable** [Categorical_Variable_Flag] Indicates if the data are categorical or not.

6. **# of thresholds/classes** [Nb_Indicators] Number of classes if the flag [Categorical_Variable_Flag] is selected, or number of threshold values for continuous attributes.

7. **Threshold Values** [Thresholds] Threshold values in ascending order, separated by spaces. That field is required only for continuous data.

8. **Marginal probabilities** [Marginal_Probabilities]

    If continuous Probability not to exceed each of the above thresholds. The entries must be monotonically increasing, and separated by spaces.

    If categorical Proportion for each category. The first entry corresponds to category 0, the second to category 1, etc. The sum of all proportions should be 1.

9. **Lower tail extrapolation** [lowerTailCdf] Parametrize the lower tail of the cumulative distribution function for continuous attributes. Input "Min" must be less than or equal ($\leq$) to the minimum of the hard data, and "omega" is the power factor.

10. **Upper tail extrapolation** [upperTailCdf] Parametrize the upper tail of the cumulative distribution function for continuous attributes. Input "Max" must

be greater than or equal ($\geq$) to the maximum of the hard data, and "omega" is the power factor.

11. **Indicator kriging type** If **Median IK** [Median_Ik_Flag] is selected, the program uses median indicator kriging to estimate the ccdf. Otherwise, if **Full IK** [Full_Ik_Flag] is selected, one IK system is solved for each threshold/class with a different variogram model.

12. **Hard Data Grid** [Hard_Data_Grid] Grid containing the conditioning hard data.

13. **Hard Data Property** [Hard_Data_Property] Conditioning data for the simulation.

14. **Assign hard data to simulation grid** [Assign_Hard_Data] If selected, the hard data are relocated onto the simulation grid. The program does not proceed if the copying fails. This option improves execution speed.

15. **Interval Data | Object** [coded_props] Grid containing the interval data. Cannot be used if **Median IK** [Median_Ik_Flag] is selected, instead use the **Full IK** [Full_Ik_Flag] option with the same variogram for all thresholds.

16. **Interval Data | Properties** [coded_grid] Properties with the interval data. These data must already be properly coded and are all found on the grid [coded_grid]. There must be [Nb_Indicators] properties selected.

17. **Max Conditioning data** [Max_Conditioning_Data] Maximum number of data to be retained in the search neighborhood.

18. **Search Ellipsoid Geometry** [Search_Ellipsoid] Parametrization of the search ellipsoid, see Section 6.4.

19. **Variogram** [Variogram] Parametrization of the indicator variograms, see Section 6.5. Only one variogram is necessary if **Median IK** [Median_Ik_Flag] is selected. Otherwise there are [Nb_Indicators] indicator variograms.

Example

The *SISIM* algorithm is run on the point-set grid shown in Fig. 4.1a. Two conditional *SISIM* realizations with a median IK regionalization are shown in Fig. 8.11. The single indicator variogram used is:

$$\gamma(h_x, h_y) = 0.07 Sph\left(\sqrt{\left(\frac{h_x}{10}\right)^2 + \left(\frac{h_y}{15}\right)^2}\right) + 0.14 Sph\left(\sqrt{\left(\frac{h_x}{40}\right)^2 + \left(\frac{h_y}{75}\right)^2}\right).$$

The search ellipsoid is of size $80 \times 80 \times 1$, with a maximum of 25 conditioning data. The lower tail extrapolation is a power model with parameter 3 and minimum value 3.4; the upper tail extrapolation is a power model with parameter 0.333 and maximum value 8.4. The ten thresholds considered and their respective cdfs are given in Table 8.1. Note that the simulated realizations exhibit more continuity of the extreme thresholds than those obtained by *SGSIM*.

Table 8.1 *Thresholds and cdfs for* SISIM *simulation*

threshold	cdf	threshold	cdf
3.5	0.0257	6.0	0.6415
4.0	0.1467	6.5	0.7830
4.5	0.2632	7.0	0.8888
5.0	0.3814	7.5	0.9601
5.5	0.5041	8.0	0.9934

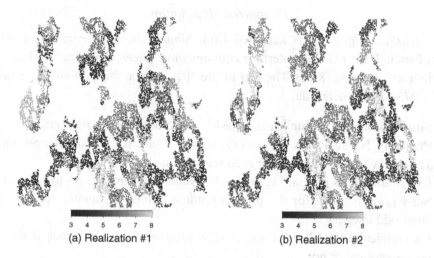

(a) Realization #1   (b) Realization #2

Figure 8.11 Two *SISIM* realizations obtained with median IK

8.1.6 COSISIM: *sequential indicator co-simulation*

Algorithm *COSISIM* extends the *SISIM* algorithm to handle secondary data. In contrast to *SISIM*, data must already be indicator-coded prior to using *COSISIM*. The algorithm does not differentiate between hard and interval data, both can be used; for any given threshold both types of data must be related to the same property. If no secondary data are selected, algorithm *COSISIM* performs a traditional sequential indicator simulation.

The secondary data are integrated using the Markov–Bayes algorithm, see Section 3.6.4, and Zhu and Journel (1993), Deutsch and Journel (1998, p.90). As with the primary attribute, the secondary information must be coded into indicators before it is used. The Markov–Bayes calibration coefficients are not internally computed and must be given as input (Zhu and Journel, 1993). Section 10.2.2 gives a Python script to calculate such coefficient values. SGeMS allows using the Markov–Bayes algorithm with both a full IK or a median IK regionalization model.

A note on conditioning

As opposed to *SISIM* where the indicator coding is done internally, *COSISIM* does not exactly honor hard data for a continuous attribute. The algorithm honors these data approximately in the sense that the simulated values are inside the correct threshold interval. It is not possible to honor exactly the original continuous data values since these were never provided to the program. A possible post-processing is to copy the conditioning hard data values over the simulated nodes once the realizations are finished.

Parameters description

The *COSISIM* algorithm is activated from *Simulation* → *cosisim* in the Algorithm Panel. The *COSISIM* interface contains three pages: "General", "Data" and "Variogram" (see Fig. 8.12). The text inside "[ ]" is the corresponding keyword in the *COSISIM* parameter file.

1. **Simulation Grid Name** [Grid_Name]   Name of the simulation grid.
2. **Property Name Prefix** [Property_Name]   Prefix for the simulation output. The suffix __real# is added for each realization.
3. **# of realizations** [Nb_Realizations]   Number of simulations to generate.
4. **Seed** [Seed]   Seed for the pseudo random number generator (preferably a large odd integer).
5. **Categorical variable** [Categorical_Variable_Flag]   Indicates if the data are categorical or not.
6. **# of thresholds/classes** [Nb_Indicators]   Number of classes if the flag [Categorical_Variable_Flag] is selected or number of threshold values for continuous attributes.
7. **Threshold Values** [Thresholds]   Threshold values in ascending order, there must be [Nb_Indicators] values entered. That field is only for continuous data.
8. **Marginal probabilities** [Marginal_Probabilities]
   If continuous Probability not to exceed each of the above thresholds. The entries must be monotonically increasing.
   If categorical Proportion for each category. The first entry corresponds to category 0, the second to category 1, ...
9. **Lower tail extrapolation** [lowerTailCdf]   Parametrize the lower tail of the ccdf for continuous attributes. Input "Min" must be less than or equal to ($\leq$) the minimum of the hard data, and "omega" is the power factor.
10. **Upper tail extrapolation** [upperTailCdf]   Parametrize the upper tail of the ccdf for continuous attributes. Input "Max" must be greater or equal to ($\geq$) the maximum of the hard data, and "omega" is the power factor.

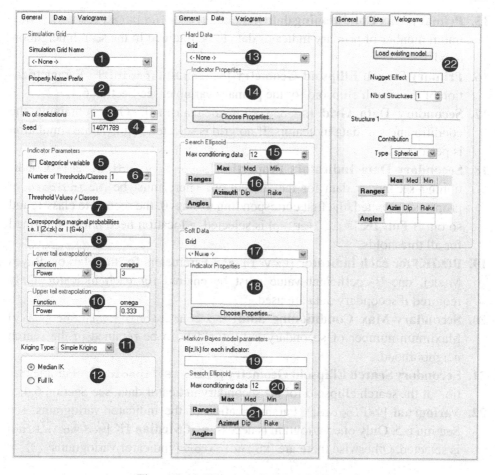

Figure 8.12 User interface for *COSISIM*

11. **Kriging Type** [Kriging_Type]   Select the type of kriging system to be solved at each node along the random path.

12. **Indicator kriging type**   If **Median IK** [Median_Ik_Flag] is selected, the program uses median indicator kriging to estimate the cdf. If **Full IK** [Full_Ik_Flag] is selected, there are [Nb_Indicators] IK systems solved at each location, one for each threshold/class.

13. **Hard Data Grid** [Primary_Hard_Data_Grid]   Grid containing the conditioning hard data.

14. **Hard Data Indicators Properties** [Primary_Indicators]   Conditioning primary data for the simulation. There must be [Nb_Indicators] properties selected, the first corresponding to class 0, the second to class 1, and so on. If **Full IK** [Full_Ik_Flag] is selected, a location may not be informed for all thresholds.

15. **Primary Max Conditioning data** [Max_Conditioning_Data_Primary] Maximum number of primary indicator data to be retained in the search neighborhood.

16. **Primary Search Ellipsoid Geometry** [Search_Ellipsoid_1] Parametrization of the search ellipsoid for the primary variable, see Section 6.4.

17. **Secondary Data Grid** [Secondary_Harddata_Grid] Grid containing the conditioning soft data indicators. If no grid is selected, a univariate simulation is performed.

18. **Secondary Data Indicators Properties** [Secondary_Indicators] Conditioning secondary data for the simulation. There must be [Nb_Indicators] properties selected, the first corresponding to class 0, the second to class 1, and so on. If **Full IK** [Full_Ik_Flag] is selected, a location need not be informed for all thresholds.

19. **B(z,IK) for each indicator** [Bz_Values] Parameters for the Markov–Bayes Model, one B-coefficient value must be entered for each indicator. Only required if secondary data are used.

20. **Secondary Max Conditioning data** [Max_Conditioning_Data_Secondary] Maximum number of secondary indicator data to be retained in the search neighborhood.

21. **Secondary Search Ellipsoid Geometry** [Search_Ellipsoid_1] Parametrization of the search ellipsoid for the secondary indicator data, see Section 6.4.

22. **Variogram** [Variogram] Parametrization of the indicator variograms, see Section 6.5. Only one variogram is necessary if **Median IK** [Median_Ik_Flag] is selected. Otherwise there are [Nb_Indicators] indicator variograms.

Example

The *COSISIM* algorithm is run on the point-set grid shown in Fig. 4.1a. Two conditional *COSISIM* realizations with a median IK regionalization are shown in Fig. 8.13. The variogram for the median indicator threshold of the primary variable is:

$$\gamma(h_x, h_y) = 0.02 + 0.23 Sph \left(\sqrt{\left(\frac{h_x}{22.5}\right)^2 + \left(\frac{h_y}{84}\right)^2} \right).$$

The search ellipsoid is of size $60 \times 60 \times 1$, with a maximum of 25 conditioning data. The lower tail extrapolation is a power model with parameter 3 and minimum value 3.4; the upper tail extrapolation is a power model with parameter 0.333 and maximum value 8.4. The ten thresholds and their respective cdfs are given in Table 8.1.

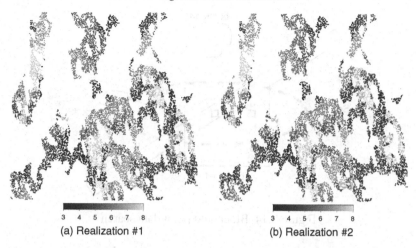

(a) Realization #1   (b) Realization #2

Figure 8.13 Two *COSISIM* realizations with median IK and the Markov–Bayes model

8.1.7 BSSIM: *block sequential simulation*

BSSIM is an algorithm for simulating point values conditioned to block and point data. The algorithm utilizes block kriging and direct sequential simulation (Hansen *et al.*, 2006; Liu, 2007).

Simulation path

There are two simulation path options in *BSSIM*. The first one is the common fully random simulation path, i.e. nodes within the simulation field are visited in random sequence. The second option is to simulate first the nodes covered by a larger number of blocks. Nodes covered by the same number of blocks have the same priority in terms of simulation sequence. This stratified random path scheme is called 'block-first' simulation path.

Block and point data search

For each simulation node, all block data are sorted based on decreasing block-to-node covariance value. Blocks with a zero block-to-node covariance are excluded from the neighborhood of this node. This block sorting is performed only once and the results are stored to be used for all realizations. The maximum number N of retained conditioning blocks is specified by the user.

To reproduce the block data, all previously simulated nodes located in a block that contains the current simulation node must be included into the point data search neighborhood, even if this leads to exceeding the specified maximum number of conditioning point data. For example, the nodes $\mathbf{u}_4$ in block #1 and $\mathbf{u}_5$ in block #2 of Fig. 8.14 should be considered for the simulation of node $\mathbf{u}_0$. However, this may result in a very large number of point conditioning data and heavy CPU demand. In *BSSIM*, three options are provided to deal with this issue. The

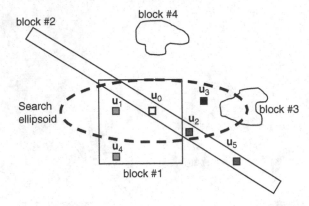

Figure 8.14 Block and point data search

first option, best but expensive for larger data sets, is to include all the nodes into the neighborhood. For example, in Fig. 8.14, in addition to the informed nodes $\mathbf{u}_1$, $\mathbf{u}_2$ and $\mathbf{u}_3$ in the search ellipsoid, informed nodes $\mathbf{u}_4$ and $\mathbf{u}_5$ are included into the conditioning data set of $\mathbf{u}_0$. The second option considers only the informed nodes $\mathbf{u}_1$, $\mathbf{u}_2$ and $\mathbf{u}_3$ within the search ellipsoid. The third option specifies the maximum number N' of the informed nodes beyond the search ellipsoid but within blocks overlapping the location to be simulated. For example, in Fig. 8.14, if N' is set to 1, $\mathbf{u}_4$ is included but not $\mathbf{u}_5$ because $\mathbf{u}_5$ has a smaller covariance value with $\mathbf{u}_0$ than $\mathbf{u}_4$.

The workflow of *BSSIM* is given in Algorithm 8.9.

Algorithm 8.9 Block sequential simulation

1: Generate the point-to-point covariance look-up table. If the FFT-integration hybrid block covariance calculation method is used, compute the block-to-point covariance map(s)

2: Define either a fully random or a block-first simulation path visiting each node **u** of the grid

3: **for** Each location **u** along the path **do**

4:     Search the conditioning data consisting of closest original point data, previously simulated values and block data

5:     Compute or retrieve the local block-to-block, block-to-point, point-to-block and point-to-point covariances

6:     Build and solve the mixed-scale kriging system.

7:     Define an appropriate local ccdf with its mean and variance given by the kriging estimate and variance

8:     Draw a value from that ccdf and add the simulated value to the data set

9: **end for**

10: Repeat to generate another simulated realization

Parameters description

The *BSSIM* algorithm is activated from *Simulation* → *bssim* in the Algorithm Panel. The *BSSIM* interface contains four pages: "General", "Data", "Variogram" and "Distribution" (see Fig. 8.15). The text inside "[ ]" is the corresponding keyword in the *BSSIM* parameter file.

1. **Simulation Grid Name** [Grid_Name]   Name of the simulation grid.
2. **Property Name Prefix** [Property_Name]   Prefix for the simulation output. The suffix _real# is added for each realization.
3. **# of realizations** [Nb_Realizations]   Number of simulated realizations to generate.
4. **Seed** [Seed]   Seed for the random number generator (preferably a large odd integer).
5. **Kriging Type** [Kriging_Type]   Select the type of kriging system to be solved at each node along the random path: **Simple Kriging (SK)** or **Ordinary Kriging (OK)**.
6. **SK mean** [SK_mean]   Mean of the attribute. Only required if **Kriging Type** [Kriging_Type] is set to **Simple Kriging(SK)**.
7. **Block Covariance Computation Approach** [Block_Cov_Approach]   Select the method of computing block covariance: **FFT with Covariance-Table** or **Integration with Covariance-Table**.
8. **Simulation Path** [Simulation_Path]   Select the simulation path scheme of visiting each node in the field: **Block First** or **Fully Random**.
9. **Check block data reproduction** [Check_Block_Reproduction]   If checked, the simulated block average values are calculated and the relative errors (compared to the input block data) are computed for each realization. The results are shown in the **Commands Panel**, which is activated from *View | Commands Panel*.
10. **Generate E-type if multiple realizations** [Generate_Etype]   If checked, an E-type is generated if there are more than one realization.
11. **Show intermediate Map (Debug)** [Debug_Info]   If set to **Show intermediate Map (Debug)**, some intermediate maps or cubes are generated, such as simulation path, point or block data neighborhood, CPU cost at some intermediate steps, etc.
12. **Hard Data | Object** [Hard_Data.grid]   Name of the grid containing the conditioning point data. If no grid is selected, the realizations are generated without conditioning point data.
13. **Hard Data | Property** [Hard_Data.property]   Property for conditioning to point data. Only required if a grid has been selected in **Hard Data | Object** [Hard_Data.grid].

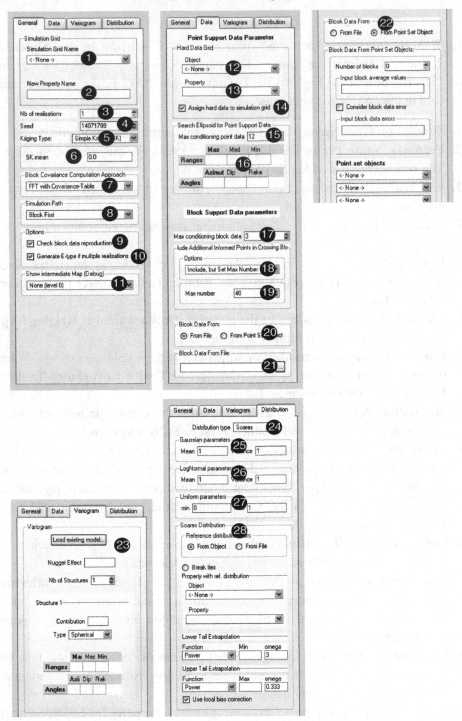

Figure 8.15 User interface for *BSSIM*

14. **Assign hard data to simulation grid** [Assign_Hard_Data] If selected, the hard data are relocated onto the simulation grid.

15. **Max conditioning point data** [Max_Conditioning_Data_Point] Maximum number of point data to be retained in the search neighborhood.

16. **Search Ellipsoid for Point Support Data** [Search_Ellipsoid_Point] Parametrization of the search ellipsoid for point support data.

17. **Max conditioning block data** [Max_Conditioning_Data_Block] Maximum number N of block data to be retained in the search neighborhood. All blocks are sorted based on their covariance with the current simulation node. Only the closest N or lesser number of blocks with non-zero block-to-current_node covariance are considered.

18. **Include Additional Informed Points in Crossing Blocks** Three options are provided here: **Include All Available** [Include All Available] corresponding to the first option in Section 8.1.7, **Not Include** [Not Include] corresponding to the second option in Section 8.1.7 and **Include, but Set Max Number** [Include, but Set Max Number] corresponding to the third option in Section 8.1.7.

19. **Max conditioning points within blocks** [Max_Cond_Points_In_Blocks] Only activated if **Include, but Set Max Number** is selected. Only the closest N' points within the crossing blocks are considered as additional conditioning data.

20. **Block Data From** Select where the block data are to be found. There are two options: **From File** [Block_From_File] and **From Point Set Object** [Block_From_Pset].

21. **Block Data From File** [Block_Data_File] Only activated if **From File** [Block_From_File] is selected in **Block Data From**. The directory address of the block data file should be specified. The block data file format is shown in Fig. 7.9. If no block data file is entered, the estimation is performed using only point data.

22. **Block Data From Point Set Objects** Only activated if **From Point Set Object** [Block_From_Pset] is selected in **Block Data From**. More details are given in Section 7.4.

23. **Variogram** [Variogram_Cov] Parametrization of the variogram model, see Section 6.5.

24. **Distribution type** [cdf_type] Select the type of ccdf to be built at each location along the random path: **Soares**, **LogNormal**, **Gaussian** and **Uniform**.

25. **Gaussian parameters** Only activated if **Distribution type** [cdf_type] is set to **Gaussian**. The parametrization of the global Gaussian distribution is done through its mean specified by **Mean** [Gaussian_mean] and its variance specified by **Variance** [Gaussian_variance].

26. **LogNormal parameters** Only activated if **Distribution type** [cdf_type] is set to **LogNormal**. The parametrization of the global lognormal distribution is done through its mean specified by **Mean** [LN_mean] and its variance specified by **Variance** [LN_variance].

27. **Uniform parameters** Only activated if **Distribution type** [cdf_type] is set to **Uniform**. Parametrization of the global Uniform distribution, the minimum is specified by **Min** [U_min] and the maximum by **Max** [U_max].

28. **Soares Distribution** [nonParamCdf] Only activated if **Distribution type** [cdf_type] is set to **Soares**. Parametrization of the global distribution from which the local distribution is sampled (see Section 6.8).

Examples

The tomography and downscaling cases used in the *BKRIG* examples (Fig. 7.11) are run using *BSSIM*.

In all these *BSSIM* runs simple kriging is used. The SK mean for the tomography case is 3.0 and that for the downscaling case is 2.8. The simulation path is fully random. The block covariance is calculated through the FFT-integration hybrid approach. All previously simulated node values within blocks containing the current simulation node are included into the data neighborhood. The variogram model (7.11) is used. The maximum number of conditioning data for point and block are both 12. The Soares method is used for histogram reproduction. The power function is used for lower tail extrapolation with parameter $\omega = 3$ and the high tail extrapolation with parameter $\omega = 0.333$.

Figure 8.16a gives a realization for the tomography data set, i.e. conditioned to both the 18-ray data and the two well data. The general patterns in the reference

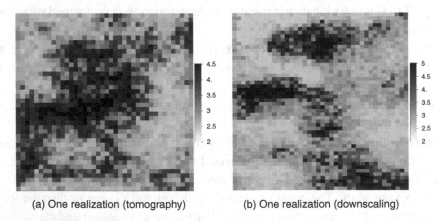

(a) One realization (tomography)   (b) One realization (downscaling)

Figure 8.16 Simulation results from *BSSIM* (tomography and downscaling cases)

model (Fig. 7.11a) are reasonably reproduced. The realization reproduces the block data with an average absolute error of 0.8%.

Figure 8.16b gives the results for the downscaling case. Again, the block data are well reproduced: the average absolute errors are 0.4%. Note that these errors are less than those for the tomography case because the tomography ray data span longer distances, exceeding the correlation range, which makes the corresponding block data reproduction difficult.

8.1.8 BESIM: *block error simulation*

BESIM (Journel and Huijbregts, 1978; Gloaguen *et al.*, 2005; Liu, 2007) is an alternative algorithm to generate stochastic realizations conditioned to both point and block data; it utilizes the approach of direct error simulation, see Section 3.8.3. This approach is extended as follows to account for block data. The kriging estimates $Z_K^*(\mathbf{u})$ and $Z_{Ks}^{*(l)}(\mathbf{u})$ in expression (3.40) must account for both point data $D_P(\mathbf{u}_\alpha)$ and block data $D_B(\mathbf{v}_\beta)$:

$$Z_K^*(\mathbf{u}) = \sum_\alpha \lambda_\alpha D_P(\mathbf{u}_\alpha) + \sum_\beta \nu_\beta D_B(\mathbf{v}_\beta)$$

$$Z_{Ks}^{*(l)}(\mathbf{u}) = \sum_\alpha \lambda_\alpha D_{Ps}(\mathbf{u}_\alpha) + \sum_\beta \nu_\beta D_{Bs}(\mathbf{v}_\beta)$$

where the λ_αs are the kriging weights for point data and the ν_βs are the kriging weights for block data.

One issue related to the second kriging estimate $Z_{Ks}^{*(l)}(\mathbf{u})$ is that the simulated conditioning block data $D_{Bs}(\mathbf{v}_\alpha)$ derived from the unconditional simulation do not contain the error component. In order to restore an error in the simulated conditioning block data, a random error value $R_s(\mathbf{v}_\alpha)$ is drawn from some distribution, e.g. a Gaussian distribution with zero mean and variance $\sigma_R^2(\mathbf{v}_\alpha)$. Hence:

$$D_{Bs}(\mathbf{v}_\alpha) = B_s(\mathbf{v}_\alpha) + R_s(\mathbf{v}_\alpha) \text{ at each block data location } \mathbf{v}_\alpha.$$

BESIM *or* BSSIM?

BESIM is a faster algorithm than *BSSIM*, see Section 3.8.3. There are two reasons. First, it only needs to solve the kriging system once per node no matter the number of realizations, while in *BSSIM*, the kriging system has to be solved once per node and per realization. Second, in *BSSIM*, all previously simulated node values within blocks containing the current simulation node are included into the data neighborhood. This may result in large kriging systems which slow down the simulation.

BESIM can not reproduce a target histogram unless that histogram is Gaussian. In all other cases, *BESIM* realizations can only approximate the target histogram through a post-processing using programs like *TRANS*, at the cost of weakening the block data conditioning, see Section 9.1 and Deutsch and Journel (1998, p.227).

The workflow of *BESIM* is given in Algorithm 8.10.

Algorithm 8.10   Block error simulation

1: Generate and store the point-to-point covariance look-up table. If the FFT-integration hybrid block covariance calculation method is used, compute and store the block-to-point covariance map(s)

2: Perform the kriging estimation for $Z_K^*(\mathbf{u})$, using the original point and block data. Store all the kriging weights for each simulation location. Note that this is only done once per location

3: Perform unconditional simulation, $Z_S(\mathbf{u})$

4: The simulated point and block data are retrieved from the unconditional simulation realization at the original data locations. Add noise drawn from the Gaussian distributions $G(0, \sigma_R^2(\mathbf{v}_\alpha))$ to the block data

5: Compute the simulated kriging estimate, $Z_{KS}^*(\mathbf{u})$, using the simulated data and the previously stored kriging weights

6: Compute the conditional simulation realization, $Z_{CS}(\mathbf{u})$

7: Repeat the process from step 3 to generate another realization

---

Parameters description

The *BESIM* algorithm is activated from *Simulation* → *besim* in the Algorithm Panel. The *BESIM* interface contains four pages: "General", "Data", "Variogram" and "Distribution" (see Fig. 8.17). The text inside "[ ]" is the corresponding keyword in the *BESIM* parameter file.

1. **Simulation Grid Name** [Grid_Name]   Name of the simulation grid.
2. **Property Name Prefix** [Property_Name]   Prefix for the simulation output. The suffix _real# is added for each realization.
3. **# of realizations** [Nb_Realizations]   Number of simulations to generate.
4. **Seed** [Seed]   Seed for the random number generator (preferably a large odd integer).
5. **Kriging Type** [Kriging_Type]   Select the type of kriging system to be solved at each node along the random path: **Simple Kriging (SK)** and **Ordinary Kriging (OK)**.
6. **SK mean** [SK_mean]   Mean of the attribute. Only required if **Kriging Type** [Kriging_Type] is set to **Simple Kriging (SK)**.

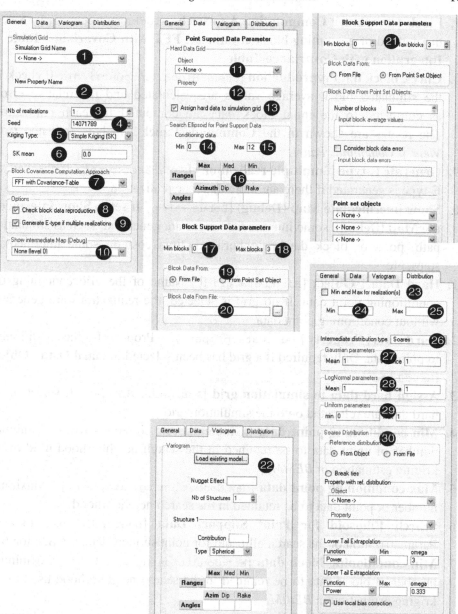

Figure 8.17 User interface for *BESIM*

7. **Block Covariance Computation Approach** [Block_Cov_Approach]   Select the method of computing block covariance: **FFT with Covariance-Table** or **Integration with Covariance-Table**.

8. **Check block data reproduction** [Check_Block_Reproduction]   If checked, the simulated block average values are calculated and the relative errors as compared to the input block data are computed for each realization. The results are shown in the **Commands Panel**, which is activated from *View | Commands Panel*.

9. **Generate E-type if multiple realizations** [Generate_Etype]   If checked, an E-type is generated if there are more than one realization.

10. **Show intermediate Map (Debug)** [Debug_Info]   If set to **Show intermediate Map (Debug)**, some intermediate maps are generated, such as simulation path, point or block data neighborhood, time cost on some intermediate steps, etc.

11. **Hard Data | Object** [Hard_Data.grid]   Name of the grid containing the conditioning point data. If no grid is selected, the realizations are generated without conditioning point data.

12. **Hard Data | Property** [Hard_Data.property]   Property for the conditioning to point data. Only required if a grid has been selected in **Hard Data | Object** [Hard_Data.grid].

13. **Assign hard data to simulation grid** [Assign_Hard_Data]   If selected, the hard data are relocated onto the simulation grid.

14. **Min conditioning point data** [Min_Conditioning_Data_Point]   Minimum number of point data to be retained in the search neighborhood used in the kriging estimation in *BESIM*.

15. **Max conditioning point data** [Max_Conditioning_Data_Point]   Maximum number of point data to be retained in the search neighborhood.

16. **Search Ellipsoid for Point Support Data** [Search_Ellipsoid_Point]   Parametrization of the search ellipsoid for point support data, see Section 6.4.

17. **Min conditioning block data** [Min_Conditioning_Data_Block]   Minimum number of block data to be retained in the search neighborhood used in the kriging estimation in *BESIM*.

18. **Max conditioning block data** [Max_Conditioning_Data_Block]   Maximum number of block data to be retained in the search neighborhood.

19. **Block Data From**   Select where the block data are defined. There are two options: **From File** [Block_From_File] and **From Point Set Object** [Block_From_Pset].

20. **Block Data From File** [Block_Data_File]   Only activated if **From File** [Block_From_File] is selected in **Block Data From**. The directory address of the block data file should be specified. The block data file format is shown in

Fig. 7.9. If no block data file is entered, the estimation is performed using only point data.

21. **Block Data From Point Set Objects** Only activated if **From Point Set Object** [Block_From_Pset] is selected in **Block Data From**. More details are given in Section 7.4.

22. **Variogram parameters** [Variogram_Cov] Parametrization of the variogram model, see Section 6.5.

23. **Min and Max for realization(s)** [Set_Realization_Min_Max] Set the minimum and maximum values for the final realization(s).

24. **Min** [Realization_Min] Only activated if **Min and Max for realization(s)** is checked. Set the minimum value in the final realization(s). All simulated values smaller than **Min** are set to **Min**.

25. **Max** [Realization_Max] Only activated if **Min and Max for realization(s)** is checked. Set the maximum value in the final realization(s). All simulated values larger than **Max** are set to **Max**.

26. **Intermediate distribution type** [cdf_type] Select the type of ccdf to be built at each location along the random path: **Soares**, **LogNormal**, **Gaussian** and **Uniform**. This is used only for the non-conditional simulation in *BESIM*. It is suggested that you use an intermediate distribution close to your target one. But histogram reproduction is not expected.

27. **Gaussian parameters** Only activated if **Distribution type** [cdf_type] is set to **Gaussian**. The parametrization of the global Gaussian distribution is done through its mean specified by **Mean** [Gaussian_mean] and its variance specified by **Variance** [Gaussian_variance].

28. **LogNormal parameters** Only activated if **Distribution type** [cdf_type] is set to **LogNormal**. The parametrization of the global lognormal distribution is done through its mean specified by **Mean** [LN_mean] and its variance specified by **Variance** [LN_variance].

29. **Uniform parameters** Only activated if **Distribution type** [cdf_type] is set to **Uniform**. Parametrization of the global Uniform distribution, the minimum is specified by **Min** [U_min] and the maximum by **Max** [U_max].

30. **Soares Distribution** [nonParamCdf] Only activated if **Distribution type** [cdf_type] is set to **Soares**. Parametrization of the global distribution from which the local distribution is sampled (see Section 6.8).

Examples

The two cases used to demonstrate *BKRIG* (Fig. 7.11) are run again using *BESIM*. Both block and point data are used for conditioning.

Figure 8.18a gives one *BESIM* realization for the tomography case. The block data are reproduced with an average absolute error of 2.6%. Figure 8.18b shows a

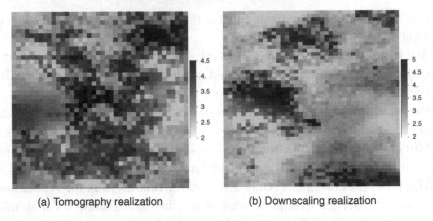

(a) Tomography realization   (b) Downscaling realization

Figure 8.18 Simulation results from *BESIM* (tomography and downscaling cases)

realization for the downscaling case. Note how the high value heterogeneity areas in the lower right and in the middle are captured. The block data reproduction,with an average absolute error of 1.0%, is better than for the above tomography case, but not as good as when using *BSSIM* (Section 8.1.7).

8.2 Multiple-point simulation algorithms

Before the introduction of multiple-point geostatistics, two families of simulation algorithms for facies modeling were available: pixel-based and object-based. The pixel-based algorithms build the simulated realizations one pixel at a time, thus providing great flexibility for conditioning to data of diverse support volumes and diverse types. Pixel-based algorithms may, however, be slow and have difficulty reproducing complex geometric shapes, particularly if simulation of these pixel values is constrained only by 2-point statistics, such as a variogram or a covariance. Object-based algorithms build the realizations by dropping onto the simulation grid one object or pattern at a time, hence they can be fast and faithful to the geometry of the object. However, they are difficult to condition to local data of different support volumes, particularly when these data are dense as in the case of seismic surveys.

SGeMS provides a program for object-based simulation *TIGENERATOR* (Section 9.9), other programs are available in many free and commercial softwares, such as "fluvsim" (Deutsch and Tran, 2002) and "SBED" (http://www.geomodeling.com).

The multiple-point simulation (mps) concept proposed by Guardiano and Srivastava (1993) and first implemented efficiently by Strebelle (2000), combines the strengths of the previous two classes of simulation algorithms. It operates pixel-wise with the conditional probabilities for each pixel value being lifted as conditional proportions from a training image depicting the geometry and distribution of objects deemed to prevail in the actual field. That training image (Ti), a purely conceptual depiction without any local accuracy, can be built using object-based algorithms. In mps the spatial patterns and structures lifted from the training image replace the 2-point structure delivered by the variogram/covariance model in traditional 2-point geostatistics. In essence, mps tries to match (exactly or approximately) a whole set of data values found in a search template/neighborhood to one or more sets of training values. The match sought is not of one datum at a time, but a match of a pattern of multiple data values taken altogether, hence the qualifier "multiple-point".

8.2.1 SNESIM: *single normal equation simulation*

The original mps implementation by Guardiano and Srivastava (1993) was much too slow, because it asked for a rescan of the whole Ti for each new multiple-point (mp) conditioning data event at each simulation node. That scan is needed to retrieve the required conditional proportion from which to draw the simulated value. The mps concept became practical with the *SNESIM* implementation of Strebelle (2000). In *SNESIM*, the Ti is scanned only once; all conditional proportions available in that Ti for a given search template size are stored in a search tree data structure, from which they can be efficiently retrieved.

The *SNESIM* algorithm contains two main parts, the construction of the search tree where all training proportions are stored, and the simulation part itself where these proportions are read and used to draw the simulated values.

Search tree construction

A search template T_J is defined by J vectors $\mathbf{h}_j$, $j = 1, \ldots, J$ radiating from a central node $\mathbf{u}_0$. The template is thus constituted by J nodes $(\mathbf{u}_0 + \mathbf{h}_j, j = 1, \ldots, J)$. That template is used to scan the training image and record all training patterns $pat(\mathbf{u}_0') = \{t(\mathbf{u}_0'); t(\mathbf{u}_0' + \mathbf{h}_j), j = 1, \ldots, J\}$, where $\mathbf{u}_0'$ is any central node of the Ti; $t(\mathbf{u}_0' + \mathbf{h}_j)$ is the training image value at grid node $\mathbf{u}_0' + \mathbf{h}_j$. All these training patterns are stored in a search tree data structure, such that one can retrieve easily the following.

1. The total number (n) of patterns with exactly the same J data values $D_J = \{d_j, j = 1, \ldots, J\}$. One such pattern would be $\{t(\mathbf{u}_0' + \mathbf{h}_j) = d_j, j = 1, \ldots, J\}$.
2. For those patterns, the number (n_k) which features a specific value $t(\mathbf{u}_0') = k$, ($k = 0, \ldots, K - 1$) at the central location $t(\mathbf{u}_0')$.

There, K is the total number of categories. The ratio of these two numbers gives the proportion of training patterns featuring the central value $t(\mathbf{u}_0') = k$ among all those identifying the J "data" values $t(\mathbf{u}_0' + \mathbf{h}_j) = d_j$:

$$P(t(\mathbf{u}_0') = k | D_J) = \frac{n_k}{n}, \quad k = 0, \ldots, K - 1. \tag{8.2}$$

Single grid simulation

The *SNESIM* sequential simulation proceeds one pixel at a time following a random path visiting all the nodes within the simulation grid G. Hard data are relocated to the closest nodes of G, and for all uninformed nodes, the classical sequential simulation (see Section 3.3.1) is used.

At each simulation node $\mathbf{u}$, the search template T_J is used to retrieve the conditional data event $dev(\mathbf{u})$ which is defined as

$$dev_J(\mathbf{u}) = \{z^{(l)}(\mathbf{u} + \mathbf{h}_1), \ldots, z^{(l)}(\mathbf{u} + \mathbf{h}_J)\}, \tag{8.3}$$

where $z^{(l)}(\mathbf{u} + \mathbf{h}_j)$ is an informed nodal value in the lth *SNESIM* realization; such value could be either an original hard datum value or a previously simulated value. Note that there can be any number of uninformed nodal values among the J possible locations of the template T_J centered at $\mathbf{u}$.

Next, find the number n of training patterns which have the same values as $dev_J(\mathbf{u})$ from the search tree. If n is less than a fixed threshold $cmin$ (the minimum number of replicates), define a smaller data event $dev_{J-1}(\mathbf{u})$ by dropping the furthest away informed node from $dev_J(\mathbf{u})$, and repeat the search. This step is repeated until $n \geq cmin$. Let J' ($J' \leq J$) be the data event size for which $n \geq cmin$.

The conditional probability, from which the nodal value $z^{(s)}(\mathbf{u})$ is drawn, is set equal to the corresponding training image proportion:

$$P(Z(\mathbf{u}) = k|dev_J(\mathbf{u})) \approx P(Z(\mathbf{u}) = k|dev_{J'}(\mathbf{u}))$$
$$= P(t(\mathbf{u}_0') = k|dev_{J'}(\mathbf{u})).$$

This probability is thus conditional to an mp data set (up to J data values) found with the search template T_J.

Algorithm 8.11 describes the simplest version of the *SNESIM* algorithm with a K-category variable $Z(\mathbf{u})$ valued in $\{0, \ldots, K - 1\}$.

The search tree stores all training replicates $\{t(\mathbf{u}_0'); t(\mathbf{u}_0' + \mathbf{h}_j), j = 1, \ldots, J\}$, and allows a fast retrieval of the conditional probability distribution of step 7 in $O(J)$. This speed comes at the cost of a possibly large RAM memory demand. Let N_{Ti} be the total number of locations in the training image. No matter the search template size J, there can not be more than N_{Ti} different data events in the training image; thus an upper-bound of the memory demand of the search tree is:

$$\text{Memory Demand} \leq \sum_{j=1}^{j=J} \min(K^j, N_{\text{Ti}})$$

where K^j is the total number of possible data value combinations with K categories and j nodes.

Algorithm 8.11 Simple single grid *SNESIM*

1: Define a search template T_J
2: Construct a search tree specific to template T_J
3: Relocate hard data to the nearest simulation grid node and freeze them during simulation
4: Define a random path visiting all locations to be simulated
5: **for** Each location **u** along the path **do**
6:    Find the conditioning data event $dev_J(\mathbf{u})$ defined by template T_J
7:    Retrieve the conditional probability distribution ccdf
     $P(Z(\mathbf{u}) = k|dev_J(\mathbf{u}))$ from search tree
8:    Draw a simulated value $z^{(s)}(\mathbf{u})$ from that conditional probability and add it to the data set
9. **end for**

Tip 2
The J number of nodes in the search template is critical for SNESIM *simulations. The greater the J value, the better the final simulated realizations, provided the training image retained is large and varied enough to provide enough replicates of such large J-point data event. However, the memory cost will increase dramatically when J is too large. For most of the 3D simulations, the J value should be set between 60 and 100, for example 80.*

Tip 3
The SNESIM *memory cost is also related to the K number of categories of the training image. In general,* SNESIM *works well with training images with no more than four categories. If the training image has more than 5 categories, then consider the hierarchical simulation approach proposed by Maharaja (2004), or use* FILTERSIM *(see Section 8.2.2).*

Multiple grid simulation

The multiple grid simulation approach (Tran, 1994) is used to capture large scale structures using a search template T_J large but with a reasonably small number of nodes. Denoted by G the 3D Cartesian grid on which simulation is to be performed, define G^g as the gth sub-set of G such that: $G^1 = G$ and G^g is obtained by down-sampling G^{g-1} by a factor of 2 along each of the 3 coordinate directions: G^g is the sub-set of G^{g-1} obtained by retaining every other node of G^{g-1}. G^g is called the gth level multi-grid. Figure 8.19 illustrates a simulation field which is divided into 3 multiple grids.

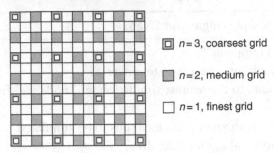

Figure 8.19 Three multiple grids (coarsest, medium and finest)

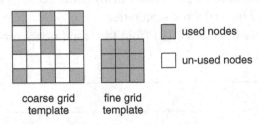

Figure 8.20 Multi-grid search template (coarse and fine)

In the gth subgrid G^g, the search template T_J is correspondingly rescaled by a factor 2^{g-1} such that

$$T_J^g = \{2^{g-1}\mathbf{h}_1, \ldots, 2^{g-1}\mathbf{h}_J\}.$$

Template T_J^g has the same number of nodes as T_J but has a greater spatial extent, hence allows capturing large-scale structures without increasing the search tree size. Figure 8.20 shows a fine template of size 3×3 and the expanded coarse template in the 2nd level coarse grid. Note that for each multiple grid a new search tree must be built.

During simulation, all nodes simulated in the previous coarser grid are frozen, i.e. they are not revisited. Algorithm 8.12 describes the implementation of multiple grids in *SNESIM*.

Anisotropic template expansion

To get the gth coarse grid, both the base search template and the simulation grid are expanded by a constant factor 2^{g-1} in all three directions. This expansion is thus "isotropic", and used as default.

The expansion factor in each direction can be made different. The gth coarse grid G^g is defined by retaining every f_x^g node, every f_y^g node and every f_z^g node in the X, Y, Z directions, respectively.

Algorithm 8.12 *SNESIM* with multiple grids

1: Choose the number N_G of multiple grids
2: Start on the coarsest grid G^g, $g = N_G$
3: **while** $g > 0$ **do**
4:    Relocate hard data to the nearest grid nodes in current multi-grid
5:    Build a new template T_j^g by re-scaling template T_J
6:    Build the search tree Tree^g using the training image and template T_J^g
7:    Simulate all nodes of G^g as in Algorithm 8.11
8:    Remove the relocated hard data from current multi-grid if $g > 1$
9:    Move to next finer grid G^{g-1} (let $g = g - 1$)
10: **end while**

The corresponding search template T_j^g is then re-scaled as:

$$T_j^g = \{\mathbf{f}^g \cdot \mathbf{h}_1, \ldots, \mathbf{f}^g \cdot \mathbf{h}_J\},$$

where $\mathbf{f}^g = \{f_x^g, f_y^g, f_z^g\}$. Note that the total number J of template nodes remains the same for all grids. This "anisotropic" expansion calls for the expansion factors to be input through the *SNESIM* interface.

Let $i = X, Y, Z$ and $1 \le g \le G$ ($G \le 10$). The requirement for the input anisotropic expansion factors are:

1. all expansion factors (f_i^g) must be positive integers;
2. expansion factor for the finest grid must be 1 ($f_i^1 \equiv 1$);
3. expansion factor for the $(g - 1)$th multi-grid must be smaller than or equal to that for the gth multi-grid ($f_i^{g-1} \le f_i^g$);
4. expansion factor for the $(g - 1)$th multi-grid must be a factor of that for the gth multi-grid ($f_i^g \bmod f_i^{g-1} = 0$).

For example, valid expansion factors for three multiple grids are:

$$
\begin{array}{ccccccc}
1\ 1\ 1 & & 1\ 1\ 1 & & 1\ 1\ 1 \\
2\ 2\ 1 & \text{or} & 4\ 2\ 2 & \text{or} & 3\ 3\ 1 & . \\
4\ 4\ 2 & & 8\ 4\ 2 & & 9\ 6\ 2
\end{array}
$$

A sensitivity analysis of the anisotropic expansion parameter should be performed before any application. An interesting development, not coded here, would be to consider a different training image for each different multiple grid.

Marginal distribution reproduction

It is sometimes desirable that the histogram of the simulated variable be close to a given target distribution, e.g. the sample histogram. There is, however, no constraint in *SNESIM* as described in Algorithm 8.11 or Algorithm 8.12 to ensure that such target distribution be reproduced. It is recommended to select a training image whose histogram is reasonably close to the target marginal proportions. *SNESIM* provides a servo system to correct the conditional distribution function read at each nodal location from the search tree (step 7 of Algorithm 8.11) to gradually gear the histogram of the up-to-now simulated values towards the target.

Let p_k^c, $k = 0, \ldots, K - 1$, denote the proportions of values in class k simulated so far, and p_k^t, $k = 0, \ldots, K - 1$, denote the target proportions. Step 7 of Algorithm 8.11 is modified as follows.

1. Compute the conditional probability distribution as originally described in step 7 of Algorithm 8.11.
2. Correct the probabilities $P\big(Z(\mathbf{u}) = k \mid dev_J(\mathbf{u})\big)$ into:

$$P^*\big(Z(\mathbf{u}) = k \mid dev_J(\mathbf{u})\big) = P\big(Z(\mathbf{u}) = k \mid dev_J(\mathbf{u})\big) + \frac{\omega}{1 - \omega} * (p_k^t - p_k^c)$$

where $\omega \in [0, 1)$ is the servosystem intensity factor. If $\omega = 0$, no correction is performed. Conversely, if $\omega \to 1$, reproducing the target distribution entirely controls the simulation process, at the risk of failing to reproduce the training image geological structures.

If $P^*\big(Z(\mathbf{u}) = k \mid dev_J(\mathbf{u})\big) \notin [0, 1]$, it is reset to the closest bound. All updated probability values are then rescaled to sum up to 1:

$$P^{**}\big(Z(\mathbf{u}) = k \mid dev_J(\mathbf{u})\big) = \frac{P^*\big(Z(\mathbf{u}) = k \mid dev_J(\mathbf{u})\big)}{\sum_{k=1}^{K} P^*\big(Z(\mathbf{u}) = k \mid dev_J(\mathbf{u})\big)}.$$

A similar procedure can be called to reproduce a given vertical proportion curve for each horizontal layer. The vertical proportion should be provided in input as a 1D property with number of nodes in X and Y directions equal to 1, and the number of nodes in the Z direction equal to that of the simulation grid. When the user inputs both a vertical proportion curve and a global target proportion, the latter is actually ignored, see Section 8.2.1 for more details about the target proportion control.

Soft data integration

Soft (secondary) data may be available to constrain the simulated realizations. The soft data are typically obtained by remote sensing techniques, such as seismic data. Often soft data provide exhaustive but low resolution information over the whole simulation grid. *SNESIM* can account for such secondary information. The soft data $Y(\mathbf{u})$ must be first calibrated into prior probability data $P\big(Z(\mathbf{u}) = k|Y(\mathbf{u})\big)$, $k = 0, \ldots, K - 1$ related to presence or absence of a certain category k centered at location $\mathbf{u}$, where K is the total number of categories.

The tau model (Section 3.10) is used to integrate probabilities coming from the soft data and the training image. The conditional distribution function of step 7 of Algorithm 8.11, $P(Z(\mathbf{u}) = k \mid dev_J(\mathbf{u}))$, is updated into $P(Z(\mathbf{u}) = k \mid dev_J(\mathbf{u}), Y(\mathbf{u}))$ as:

$$P(Z(\mathbf{u}) = k \mid dev_J(\mathbf{u}), Y(\mathbf{u})) = \frac{1}{1+x}, \tag{8.4}$$

where the distance x is calculated as

$$\frac{x}{x_0} = \left(\frac{x_1}{x_0}\right)^{\tau_1} \left(\frac{x_2}{x_0}\right)^{\tau_2}, \qquad \tau_1, \tau_2 \in (-\infty, +\infty), \tag{8.5}$$

where the distances x_0, x_1, x_2 are defined as:

$$x_0 = \frac{1 - P(Z(\mathbf{u}) = k)}{P(Z(\mathbf{u}) = k)}$$

$$x_1 = \frac{1 - P(Z(\mathbf{u}) = k \mid dev_J(\mathbf{u}))}{P(Z(\mathbf{u}) = k \mid dev_J(\mathbf{u}))}$$

$$x_2 = \frac{1 - P(Z(\mathbf{u}) = k \mid Y(\mathbf{u}))}{P(Z(\mathbf{u}) = k \mid Y(\mathbf{u}))}.$$

$P(Z(\mathbf{u}) = k)$ is the target marginal proportion of category k. The two weights τ_1 and τ_2 account for the redundancy (Krishnan *et al.*, 2005) between the local conditioning data event $dev_J(\mathbf{u})$ and the soft data $Y(\mathbf{u})$. The default values are $\tau_1 = \tau_2 = 1$, corresponding to non-redundant data. See Section 3.10 for more details about the tau model.

Step 7 of Algorithm 8.11 is then modified as follows.

1. Estimate the probability $P(Z(\mathbf{u}) = k \mid dev_J(\mathbf{u}))$ as described in Algorithm 8.11.
2. Compute the updated probability $P(Z(\mathbf{u}) = k \mid dev_J(\mathbf{u}), Y(\mathbf{u}))$ using Eq. (8.4).
3. Draw a realization from the updated distribution function.

Subgrid concept

As described earlier in Section 8.2.1, whenever *SNESIM* cannot find enough training replicates of a given data event dev_J, it will drop the furthest node in dev_J and repeat searching until the number of replicates is greater than or equal ($\geq$) to $cmin$. This data dropping procedure not only decreases the quality of pattern reproduction, but also significantly increases CPU cost.

The subgrid concept is proposed to alleviate the data dropping effect. Figure 8.21a shows the eight contiguous nodes of a 3D simulation grid, also seen

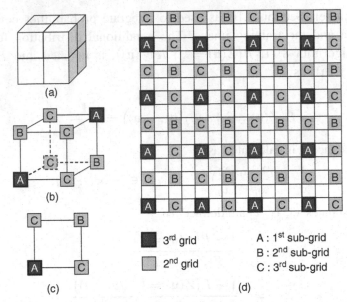

Figure 8.21   Subgrid concept: (a) 8 close nodes in 3D grid; (b) 8 nodes represented in the corners of a cubic; (c) 4 close nodes in 2D grid; (d) 3 subgrids in the 2nd multi-grid.

as the 8 corners of a cube (Fig. 8.21b). Among them, nodes 1 and 8 belong to subgrid 1; nodes 4 and 5 belong to subgrid 2, and all other nodes belong to subgrid 3. Figure 8.21c shows the subgrid concept in 2 dimensions. The simulation is performed first over subgrid 1, then subgrid 2 and finally over subgrid 3. This subgrid simulation concept is applied to all multiple grids except the coarsest grid. Figure 8.21d shows the 3 subgrids over the 2nd multi-grid, where "A" denotes the 1st subgrid nodes, "B" denotes the 2nd subgrid nodes, and "C" denotes the 3rd subgrid nodes.

In the 1st subgrid of the gth multi-grid, most of the nodes (of type A) would have been already simulated in the previous coarse $(g + 1)$th multi-grid: with the default isotropic expansion, 80% of these nodes are already simulated in the previous coarser grid in 3D and 100% of them in 2D. In that 1st subgrid, the search template is designed to only use the A-type nodes as conditioning data, hence the data event is almost full. Recall that the nodes previously simulated in the coarse grid are not resimulated in this subgrid.

In the 2nd subgrid, all the nodes marked as "A" in Fig. 8.21d are now informed by a simulated value. In this subgrid, the search template T_J is designed to use only the A-type nodes, but conditioning includes in addition the J' closest B-type nodes; the default is $J' = 4$. In total, there are $J + J'$ nodes in the search template T for that 2nd subgrid. The left plot of Fig. 8.22 shows the 2nd subgrid nodes and the template nodes of a simple 2D case with isotropic expansion: the basic search template of size 14 is marked by the solid circles, the $J' = 4$ additional conditioning

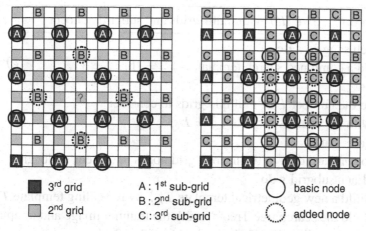

■	3rd grid	A : 1st sub-grid	◯	basic node
■	2nd grid	B : 2nd sub-grid	⬭	added node
		C : 3rd sub-grid		

Figure 8.22 Simulation nodes and search template in subgrid (left: 2nd subgrid; right: 3rd subgrid)

nodes are marked by the dash circles. Note that the data event captured by the basic template nodes (solid circles) is always full.

When simulating over the 3rd subgrid, all nodes in both the 1st and the 2nd subgrids (of types A and B) are fully informed with simulated values. In that 3rd subgrid, the base template T_J is designed to search only nodes of types A and B for a conditioning data event informed as either original hard data or previously simulated data. Again, J' nearest nodes of type C in the current subgrid are used as additional conditioning data. In the right plot of Fig. 8.22, the basic search template for the 3rd subgrid is marked by solid circles, the $J' = 4$ additional conditioning nodes are marked by dash circles.

The subgrid approach mimics a staggered grid, allowing more conditioning data to be found during simulation in each subgrid; also the search for these data is much faster. It is *strongly* recommended to use this subgrid option for 3D simulation. Algorithm 8.12 is modified as shown in Algorithm 8.13.

Tip 4

If the user wants to use the subgrid concept and the anisotropic template expansion simultaneously, all expansion factors should be power 2 value, i.e. $f_i^g = 2^n$, where $i = X, Y, Z$ and g is the coarse grid number. Otherwise, the subgrid option will be turned off automatically, because the SNESIM cannot mimic a staggered grid any more.

Node re-simulation

The other solution for reducing data dropping effect is to re-simulate those nodes that were simulated with too small a number of conditioning data, i.e. less than a

---

Algorithm 8.13   *SNESIM* with multi-grids and subgrid concept

---

1: **for** Each subgrid s **do**
2:      Build a combined search template $T^s_{j,j'} = \{\mathbf{h}_i, \; i = 1, \ldots, \ldots, (J + J')\}$
3: **end for**
4: Choose the number L of multiple grids to consider
5: Start with the coarsest grid G^g, $g = L$
6: **while** $g > 0$ **do**
7:      Relocate hard data to the nearest grid nodes in current multi-grid
8:      **for** Each subgrid s **do**
9:          Build a new geometrical template $T^{g,s}_{j,j'}$ by re-scaling template $T^s_{j,j'}$
10:         Build the search tree $\text{Tree}^{g,s}$ using the training image and template $T^g_{j,j'}$
11:         Simulate all nodes of $G^{g,s}$ as in Algorithm 8.11
12:     **end for**
13:     Remove the relocated hard data from current multi-grid if $g > 1$
14:     Move to next finer grid G^{g-1} (let $g = g - 1$)
15: **end while**

---

given threshold. *SNESIM* records the number N_{drop} of data event nodes dropped during simulation. After the simulation on each subgrid of each multiple grid, the simulated values at those nodes with N_{drop} larger than a threshold are de-allocated, and pooled into a new random path. Then *SNESIM* is performed again along this new random path. This post-processing technique (Remy, 2001) improves reproduction of the large-scale patterns.

In the current *SNESIM* coding, the threshold value is input through the graphic interface. *SNESIM* also allows to repeat this post-processing through multiple iterations.

Accounting for local non-stationarity

Any training image should be reasonably stationary so that meaningful statistics can be inferred by scanning it. It is however possible to introduce some non-stationarity in the simulation through local rotation and local scaling of an otherwise stationary Ti. *SNESIM* provides two approaches to handle such non-stationary simulation: (1) modify locally the training image; (2) use different training images. The first approach is presented in this section; the second will be detailed in Section 8.2.1.

The simulation field G can be divided into several rotation regions, each region associated with a rotation angle. Let r^i ($i = 0, \ldots, N_{\text{rot}} - 1$) be the rotation angle about the (vertical) Z-axis in the ith region R^i, where N_{rot} is the total number of regions for rotation, and $R^0 \cup \cdots \cup R^{N_{\text{rot}}-1} = G$. In the present *SNESIM* version one single azimuth rotation is allowed around the Z-axis, with the angle measured in degree increasing clockwise from the Y-axis. If needed, additional rotations of both the data sets and the simulation grids could be performed prior to using *SNESIM*.

The simulation grid G can also be divided into a set of scaling regions, each region associated with scaling factors in X/Y/Z directions. Let $\mathbf{f}^j = \{f_x^j, f_y^j, f_z^j\}$ ($j = 0, \ldots, N_{\text{aff}} - 1$) be the scaling factors, also called affinity ratios, in the jth region S^j, where N_{aff} is number of regions for scaling, $S^0 \cup \cdots \cup S^{N_{\text{aff}}-1} = G$, and f_x^j, f_y^j, f_z^j are the affinity factors in the X/Y/Z directions, respectively. All affinity factors must be positive $\in (0, +\infty)$. The larger the affinity factor, the larger the geological structure in that direction. An affinity factor equal to 1 means no training image scaling.

The N_{rot} rotation regions and N_{aff} scaling regions can be independent from one another, thus allowing overlap of rotation regions with scaling regions.

Given N_{rot} rotation regions and N_{aff} affinity regions, the total number of new training images after scaling and rotation is $N_{\text{rot}} \cdot N_{\text{aff}}$. Correspondingly, one different search tree must be constructed using template T_J for each new training image $\text{Ti}_{i,j}$ defined as:

$$\text{Ti}_{i,j}(\mathbf{u}) = \Theta_i \cdot \Lambda_j \cdot Ti(\mathbf{u}),$$

where $\mathbf{u}$ is the node in the training image, Θ_i is the rotation matrix for rotation region i, and Λ_j is the scaling matrix for affinity region j:

$$\Theta_i = \begin{bmatrix} \cos r^i & \sin r^i & 0 \\ -\sin r^i & \cos r^i & 0 \\ 0 & 0 & 1 \end{bmatrix} \qquad \Lambda_j = \begin{bmatrix} f_x^j & 0 & 0 \\ 0 & f_y^j & 0 \\ 0 & 0 & f_z^j \end{bmatrix}. \tag{8.6}$$

$N_{\text{rot}} = 1$ means to rotate the Ti globally, which can be also achieved by specifying one global rotation angle (see the latter *parameter description* section). Similarly $N_{\text{aff}} = 1$ corresponds to a global scaling of the Ti.

The corresponding *SNESIM* algorithm is described in Algorithm 8.14.

This solution can be very memory demanding as one new search tree has to be built for each pair of scaling factor $\mathbf{f}_j$ and rotation angle r_i. Practice has shown that it is possible to generate fairly complex models using a limited number of regions: a maximum of five rotation regions and five affinity regions is sufficient in most cases with a reasonable size training image.

Region concept

The rotation and affinity concepts presented in the previous section allow to account only for limited non-stationarity in that the geological structures in the different subdomains are similar except for orientation and size. In more difficult cases, the geological structures may be fundamentally different from one zone or region to another, calling for different training images in different regions, see R1, R2 and R3 in Fig. 8.23. Also, parts of the study field may be inactive (R4

Algorithm 8.14   *SNESIM* with locally varying azimuth and affinity

---

1: Define a search template T_J
2: **for** Each rotation region i **do**
3:    **for** Each affinity region j **do**
4:       Construct a search tree $\text{Tr}_{i,j}$ for training image $\text{Ti}_{i,j}$ using template T_J
5:    **end for**
6: **end for**
7: Relocate hard data to the nearest simulation grid nodes
8: Define a random path visiting all locations to be simulated
9: **for** Each location **u** along the path **do**
10:    Find the conditioning data event $dev_J(\mathbf{u})$ defined by template T_J
11:    Locate the region index (i, j) of location **u**
12:    Retrieve the conditional probability distribution ccdf $P(Z(\mathbf{u}) = k|dev_J(\mathbf{u}))$
       from the corresponding search tree $\text{Tr}_{i,j}$
13:    Draw a simulated value $z^{(s)}(\mathbf{u})$ from that conditional distribution and add it
       to the data set
14: **end for**

---

in Fig. 8.23), hence there is no need to perform *SNESIM* simulation in those locations, and the target proportion should be limited to only the active cells. There may be a slow transition between these subdomains, calling for data sharing at borders. The region concept allows for such gradual transitions as opposed to a set of totally independent simulations.

The simulation grid G is first divided into a set of subdomains (regions) G^i, $i = 0, \ldots, N_R - 1$, where N_R is the total number of regions, and $G^0 \cup \cdots \cup G^{N_R-1} = G$. Perform normal *SNESIM* simulation for each active region with its specific training image and its own parameter settings. The regions can be simulated in any order, or can be simulated simultaneously through a random path visiting all regions.

To simulate a domain with regions, the user has to simulate each region sequentially. Except for the first region, the simulations in any region are conditioned to close-by values simulated in other regions, in order to reduce discontinuity across region boundaries. The simulated result contains not only the property values in the current region, but also the property copied from the other conditioning regions. For instance in Fig. 8.23, when region 2 (R2) is simulated conditional to the property re_1 in region 1 (R1), the simulated realization $re_{1,2}$ contains the property in both R1 and R2. Next the property $re_{1,2}$ can be used as conditioning data to perform *SNESIM* simulation in region 3 (R3), which will result in a realization over all the active areas (R1 + R2 + R3).

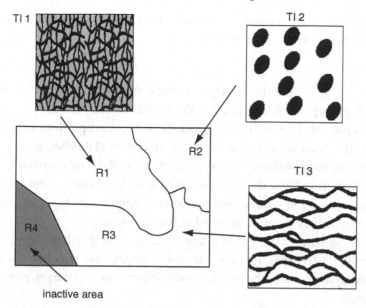

Figure 8.23 Simulation with region concept: each region is associated to a specific Ti

Tip 5

Although the region concept has to be applied manually, advanced users can use Python scripts to automate the simulation tasks (see details in Section 10.2). Because the regions are processed in sequence, only one search tree is actually saved in memory during the simulation. This significantly reduces the memory cost. Hence with the Python scripts, the user can account for local non-stationary constraints with more rotation and scaling regions, for instance $N_{rot} \times N_{aff} > 25$.

Target distributions

SNESIM allows three types of target proportions: a global target proportion, a vertical proportion curve and a soft probability cube. Three indicators I_1, I_2, I_3 are defined as follows.

$$I_1 = \begin{cases} 1 & \text{a global target is given} \\ 0 & \text{no global target} \end{cases}$$

$$I_2 = \begin{cases} 1 & \text{a vertical proportion curve is given} \\ 0 & \text{no vertical proportion curve} \end{cases}$$

$$I_3 = \begin{cases} 1 & \text{a probability cube is given} \\ 0 & \text{no probability cube} \end{cases}$$

There is a total $2^3 = 8$ possible options. The *SNESIM* program will proceed as follows, according to the given option.

1. $I_1 = I_2 = I_3 = 1$ [global target, vertical proportion curve and probability cube all given] *SNESIM* ignores the global target, and checks consistency between the soft probability cube and the vertical proportion. If these are not consistent, a warning is prompted in the algorithm status bar, and the program continues running without waiting for correction of the inconsistency. The local conditional probability distribution (ccdf) is updated first for the soft probability cube using a tau model, then the servosystem is enacted using the vertical probability values as target proportion for each layer.

2. $I_1 = I_2 = 1, I_3 = 0$ [global target, vertical proportion curve, no probability cube] *SNESIM* ignores the global target, and corrects the ccdf with the servosystem using the vertical probability value as target proportion for each layer.

3. $I_1 = 1, I_2 = 0, I_3 = 1$ [global target, no vertical proportion curve, probability cube] *SNESIM* checks the consistency between the soft probability cube and the global target proportion. If these are not consistent, a warning is prompted in the algorithm status bar and the program continues running without correcting the inconsistency. The ccdf is updated first for the soft probability cube using the tau model, then the servosystem is enacted to approach the global target proportion.

4. $I_1 = 1, I_2 = I_3 = 0$ [global target, no vertical proportion curve, no probability cube] *SNESIM* corrects the ccdf with the servosystem to approach the global target proportion.

5. $I_1 = 0, I_2 = I_3 = 1$ [no global target, vertical proportion curve, probability cube] Same as case 1.

6. $I_1 = 0, I_2 = 1, I_3 = 0$ [no global target, vertical proportion curve, no probability cube] Same as case 2.

7. $I_1 = I_2 = 0, I_3 = 1$ [no global target, no vertical proportion curve, probability cube] *SNESIM* gets the target proportion from the training image, then checks the consistency between the soft probability cube and that target proportion. If these are not consistent, a warning is prompted in the algorithm status bar and the program continues running without correcting the inconsistency. The ccdf is updated first for the soft probability cube using a tau model, then the servosystem is enacted to approach the target proportion.

8. $I_1 = I_2 = I_3 = 0$ [no global target, no vertical proportion curve, no probability cube] *SNESIM* gets the target proportion from the training image, then corrects the ccdf with the servosystem to approach the global proportion.

Parameters description

The *SNESIM* algorithm is activated from *Simulation* → *snesim_std* in the upper part of the Algorithm Panel. The main *SNESIM* interface contains four pages: "General", "Conditioning", "Rotation/Affinity" and "Advanced" (see Fig. 8.24). The *SNESIM* parameters will be presented page by page in the following. The text inside "[ ]" is the corresponding keyword in the *SNESIM* parameter file.

1. **Simulation Grid Name** [GridSelector_Sim]   The name of grid on which simulation is to be performed.
2. **Property Name Prefix** [Property_Name_Sim]   Prefix for the simulation output. The suffix __real# is added for each realization.
3. **# of Realizations** [Nb_Realizations]   Number of realizations to be simulated.
4. **Seed** [Seed]   A large odd integer to initialize the pseudo-random number generator.
5. **Training Image | Object** [PropertySelector_Training.grid]   The name of the grid containing the training image.
6. **Training Image | Property** [PropertySelector_Training.property]   The training image property, which must be a categorical variable whose value must be between 0 and $K - 1$, where K is the number of categories.
7. **# of Categories** [Nb_Facies]   The number K of categories contained in the training image.
8. **Target Marginal Distribution** [Marginal_Cdf]   The target category proportions, must be given in sequence from category 0 to category Nb_Facies-1. The sum of all target proportions must be 1.
9. **# of Nodes in Search Template** [Max_Cond]   The maximum number J of nodes contained in the search template. The larger the J value, the better the simulation quality if the training image is correspondingly large, but the more demand on RAM memory. Usually, around 60 nodes in 2D and 80 nodes in 3D with multigrid option should create fairly good realizations.
10. **Search Template Geometry** [Search_Ellipsoid]   The ranges and angles defining the ellipsoid used to search for neighboring conditioning data. The search template T_J is automatically built from the search ellipsoid retaining the J closest nodes.
11. **Hard Data | Object** [Hard_Data.grid]   The grid containing the hard conditioning data. The hard data object must be a point set. The default input is "None", which means no hard conditioning data are used.
12. **Hard Data | Property** [Hard_Data.property]   The property of the hard conditioning data, which must be a categorical variable with values between 0 and $K - 1$. This parameter is ignored when no hard data conditioning is selected.

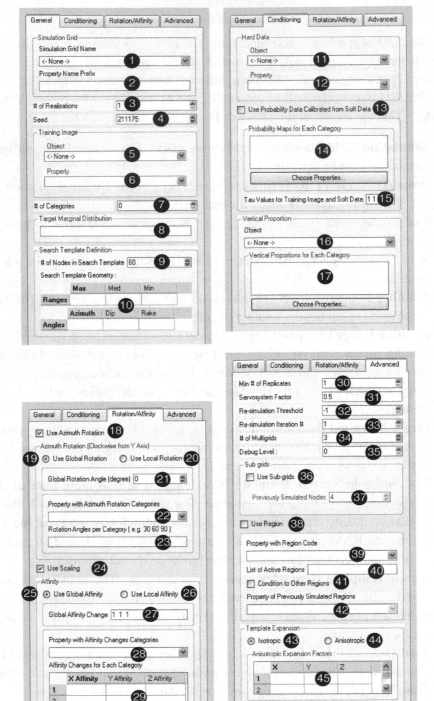

Figure 8.24 *SNESIM* interface

13. **Use Probability Data Calibrated from Soft Data** [Use_ProbField] This flag indicates whether the simulation should be conditioned to prior local probability cubes. If marked, perform *SNESIM* conditional to prior local probability information. The default is not to use soft probability cubes.

14. **Soft Data | Choose Properties** [ProbField_properties] Selection for the soft probability data. One property must be specified for each category k. The property sequence is critical to the simulation result: the kth property corresponds to $P(Z(\mathbf{u}) = k \mid Y(\mathbf{u}))$. This parameter is ignored if Use_ProbField is set to 0. Note that the soft probability data must be given over the same simulation grid defined in (1).

15. **Tau Values for Training Image and Soft Data** [TauModelObject] Input *two* tau parameter values: the first tau value is for the training image, the second tau value is for the soft conditioning data. The default tau values are "1 1". This parameter is ignored if Use_ProbField is set to 0.

16. **Vertical Proportion | Object** [VerticalPropObject] The grid containing the vertical proportion curve. This grid must be 1D: the number of cells in the X and Y directions must be 1, and the number of cells in the Z direction must be the same as that of the simulation grid. The default input is "None", which means no vertical proportion data is used.

17. **Vertical Proportion | Choose Properties** [VerticalProperties] Select one and only one proportion for each category k. The property sequence is critical to the simulation result. If VerticalPropObject is "None", then this parameter is ignored.

18. **Use Azimuth Rotation** [Use_Rotation] The flag to use the azimuth rotation concept to handle non-stationary simulations. If marked (set as 1), then use rotation concept. The default is not to use rotation.

19. **Use Global Rotation** [Use_Global_Rotation] To rotate the training image with a single azimuth angle. If marked (set as 1), a single angle must be specified in "Global Rotation Angle".

20. **Use Local Rotation** [Use_Local_Rotation] To rotate the training image for each region. If selected, a rotation angle must be specified for each region in Rotation_ categories. Note that Use_Global_Rotation and Use_Local_Rotation are mutually exclusive.

21. **Global Rotation Angle** [Global_Angle] The global azimuth rotation angle given in degrees. The training image will be rotated clockwise by that angle prior to simulation. This parameter is ignored if Use_Global_Rotation is set to 0.

22. **Property with Azimuth Rotation Categories** [Rotation_property] The property containing the coding of the rotation regions, must be given over the same simulation grid as defined in (1). The region code ranges from 0

to $N_{rot} - 1$ where N_{rot} is the total number of regions. The angles corresponding to all the regions are specified by Rotation_categories.

23. **Rotation Angles per Category** [Rotation_categories] The angles, expressed in degrees, corresponding to each region. The angles must be given in sequence separated by a space. If Use_Global_Rotation is set to 0, then this parameter is ignored.

24. **Use Scaling** [Use_Affinity] The flag to use the affinity concept to handle non-stationary simulations. If marked (set as 1), use the affinity concept. The default is not to use scaling.

25. **Use Global Affinity** [Use_Global_Affinity] The flag to indicate whether to scale the training image with the same constant factor in each X/Y/Z direction. If checked (set as 1), three affinity values must be specified in "Global Affinity Change".

26. **Use Local Affinity** [Use_Local_Affinity] To scale the training image for each affinity region. If set to 1, three affinity factors must be specified for each region. Note that Use_Global_Affinity and Use_Local_Affinity are mutually exclusive.

27. **Global Affinity Change** [Global_Affinity] Input three values (separated by spaces) for the X/Y/Z directions, respectively. If the affinity value in a certain direction is f, then the category width in that direction is f times the original width; the larger f the wider the simulated bodies.

28. **Property with Affinity Changes Categories** [Affinity_property] The property containing the coding of the affinity regions, must be given over the same simulation grid as defined in (1). The region code ranges from 0 to $N_{aff} - 1$ where N_{aff} is the total number of affinity regions. The affinity factors should be specified by Affinity_categories.

29. **Affinity Changes for Each Category** [Affinity_categories] Input the affinity factors in the table: one scaling factor for each X/Y/Z direction and for each region. The region index (the first column in the table) is actually the region indicator plus 1.

30. **Min # of Replicates** [Cmin] The minimum number of training replicates of a given conditioning data event to be found in the search tree before retrieving its conditional probability. The default value is 1.

31. **Servosystem Factor** [Constraint_Marginal_ADVANCED] A parameter ($\in$ [0, 1]) which controls the servosystem correction. The higher the servosystem parameter value, the better the reproduction of the target category proportions. The default value is 0.5.

32. **Re-simulation Threshold** [resimulation_criterion] The threshold value needed for re-simulation. Those simulated nodes with N_{drop} (number of

conditioning nodes dropped during simulation) larger than the input threshold value are re-simulated. The default value is -1, which means no re-simulation.

33. **Re-simulation Iteration #** [resimulation_iteration_nb] The number of iterations to repeat the above post-processing procedure. This parameter is ignored when resimulation_criterion is -1. The default value is 1.

34. **# of Multigrids** [Nb_Multigrids_ADVANCED] The number of multiple grids to consider in the multiple grid simulation. The default value is 3.

35. **Debug Level** [Debug_Level] The option controls the output in the simulation grid. The larger the debug level, the more outputs from *SNESIM*:
    - if 0, then only the final simulation result is output (default value);
    - if 1, then a map showing the number of nodes dropped during simulation is also output;
    - if 2, then intermediate simulation results are output in addition to the outputs from options 0 and 1.

36. **Use subgrids** [Subgrid_choice] The flag to divide the simulation nodes on the current multi-grid into three groups to be simulated in sequence. It is *strongly* recommended to use this option for 3D simulation.

37. **Previously simulated nodes** [Previously_simulated] The number of nodes in current subgrid to be used for data conditioning. The default value is 4. This parameter is ignored if Subgrid_choice is set to 0.

38. **Use Region** [Use_Region] The flag indicates whether to use the region concept. If marked (set as 1), perform *SNESIM* simulation with the region concept; otherwise perform *SNESIM* simulation over the whole grid.

39. **Property with Region Code** [Region_Indicator_Prop] The property containing the index coding of the regions, must be given over the same simulation grid as defined in (1). The region code ranges from 0 to $N_R - 1$ where N_R is the total number of regions.

40. **List of Active Regions** [Active_Region_Code] Input the index of the region to be simulated, or indices of the regions to be simulated simultaneously. If simulation with multiple regions, the input region indices (codes) should be separated by spaces.

41. **Condition to Other Regions** [Use_Previous_Simulation] The option to perform region simulation conditional to data from other regions.

42. **Property of Previously Simulated Regions** [Previous_Simulation_Pro] The property simulated in the other regions. The property can be different from one region to another. See Section 8.2.1.

43. **Isotropic Expansion** [expand_isotropic] The flag to use isotropic expansion method for generating the series of cascaded search templates and multiple grids.

44. Anisotropic Expansion [expand_anisotropic] The flag to use anisotropic factors for generating a series of cascaded search templates and multiple grids.

45. Anisotropic Expansion Factors [aniso_factor] Input an integer expansion factor for each X/Y/Z direction and for each multiple grid in the given table. The first column of the table indicates the multiple grid level; the smaller the number, the finer the grid. This option is not recommended to beginners.

Examples

This section presents four examples showing how *SNESIM* algorithm works with categorical training images with or without data conditioning for both 2D and 3D simulations.

EXAMPLE 1: 2D unconditional simulation

Figure 8.25a shows a channel training image of size 150 × 150. This training image contains four facies: mud background, sand channel, levee and crevasse. The facies proportions are 0.45, 0.2, 0.2 and 0.15 respectively. An unconditional *SNESIM* simulation is performed with this training image using a maximum of 60 conditioning data, and a servosystem 0.5. The search template is isotropic in 2D with isotropic template expansion. Four multiple grids are used to capture large scale channel structures. Figure 8.25b gives one *SNESIM* realization, whose facies proportions are 0.44, 0.19, 0.2 and 0.17 respectively. It is seen that the channel continuities and the facies attachment sequence are reasonably well reproduced.

EXAMPLE 2: 3D simulation conditioning to well data and soft seismic data

In this example, the large 3D training image (Fig. 8.26a) is created with the object-based program "fluvsim" (Deutsch and Tran, 2002). The dimension of

 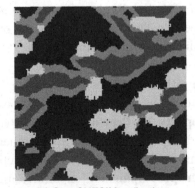

(a) Four categories training image   (b) One *SNESIM* realization

Figure 8.25 [Example 1] Four facies training image and one *SNESIM* simulation (black: mud facies; dark gray: channel; light gray: levee; white: crevasse)

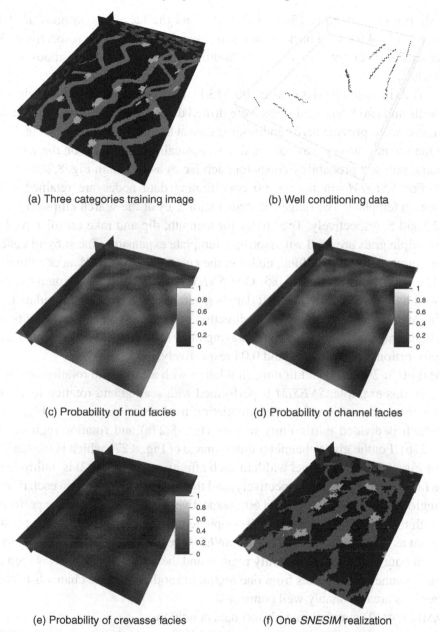

(a) Three categories training image

(b) Well conditioning data

(c) Probability of mud facies

(d) Probability of channel facies

(e) Probability of crevasse facies

(f) One *SNESIM* realization

Figure 8.26 [Example 2] Three facies 3D training image (black: mud facies; gray: channel; white: crevasse), well hard data, facies probability cubes and one *SNESIM* realization. Graphs (c)–(f) are given at the same slices: X = 12, Y = 113, Z = 4

this training image is $150 \times 195 \times 30$, and the facies proportions are 0.66, 0.3 and 0.04 for mud background, sand channel and crevasse respectively. The channels are oriented in the North–South direction with varying sinuosities and widths.

The simulated field is of size $100 \times 130 \times 10$. Two vertical wells, five deviated wells and two horizontal wells were drilled during the early production period. Those wells provide hard conditioning data at the well locations, see Fig. 8.26b. One seismic survey was collected, and was calibrated, based on the well hard data, into soft probability cubes for each facies as shown in Fig. 8.26c–e.

For *SNESIM* simulation, 60 conditioning data nodes are retained in the search template. The ranges of three major axes of the search ellipsoid are 20, 20 and 5, respectively. The angles for azimuth, dip and rake are all zero. Four multiple grids are used with isotropic template expansion. The subgrid concept is adopted with 4 additional nodes in the current subgrid for data conditioning. The servosystem value is 0.5. One *SNESIM* realization conditioning to both well hard data and seismic soft data is given in Fig. 8.26f. This simulated field has channels oriented in the NS direction, with the high sand probability area (light gray to white in Fig. 8.26d) having more sand facies. The simulated facies proportions are 0.64, 0.32 and 0.04 respectively.

EXAMPLE 3: 2D hard conditioning simulation with affinity and rotation regions

In this example, *SNESIM* is performed with scaling and rotation to account for local non-stationarity. The simulation field is the last layer of Fig. 8.26b, which is divided into affinity regions (Fig. 8.27a) and rotation regions (Fig. 8.27b). For the given channel training image of Fig. 8.27c which is the 4th layer of Fig. 8.26a, the channel width in each affinity region (0, 1, 2) is multiplied by a factor of 2, 1 and 0.5, respectively; and the channel orientation in each rotation region (0, 1, 2) is $0°$, $-60°$ and $60°$, respectively. The simulations are performed with three multiple grids using isotropic template expansion. The servosystem is set as 0.5. Figure 8.27d gives one *SNESIM* realization conditioned only to the well data using both the affinity regions and the rotation regions. It is seen that the channel width varies from one region to another; and the channels between regions are reasonably well connected.

EXAMPLE 4: 2D simulation with soft data conditioning

In this last example, the simulation grid is again the last layer of Fig. 8.26b. Both the soft data and well hard data from that layer are used for data conditioning. Figure 8.28a gives the mud probability field. Figure 8.27c is used as the training image. The search template is isotropic with 60 conditioning nodes. Four multiple grids are retained with isotropic template expansion. *SNESIM* is run for 100 realizations. Figure 8.28c–e present three realizations: the channels are well connected in the NS direction; and their positions are consistent with the soft probability data (see the dark area in Fig. 8.28a for the channel

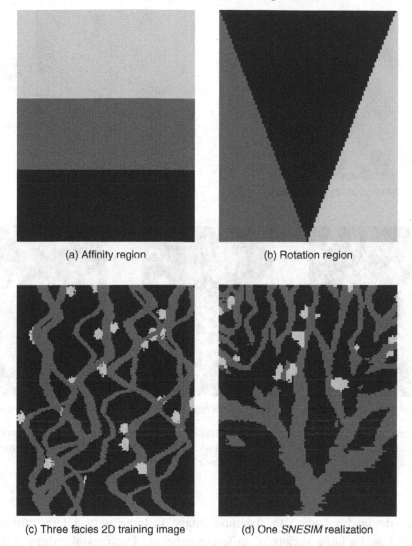

(a) Affinity region

(b) Rotation region

(c) Three facies 2D training image

(d) One *SNESIM* realization

Figure 8.27 [Example 3] Affinity and rotation regions (black: region 0; gray: region 1; white: region 2); three facies 2D training image and one *SNESIM* simulation with both affinity and rotation regions (black: mud facies; gray: sand channel; white: crevasse)

locations). Figure 8.28b gives the experimental mud facies probability obtained from the simulated 100 realizations, this experimental probability is consistent with the input mud probability Fig. 8.28a.

8.2.2 FILTERSIM: *filter-based simulation*

SNESIM is designed for modeling categories, e.g. facies distributions. It is limited in the number of categorical variables it can handle. *SNESIM* is

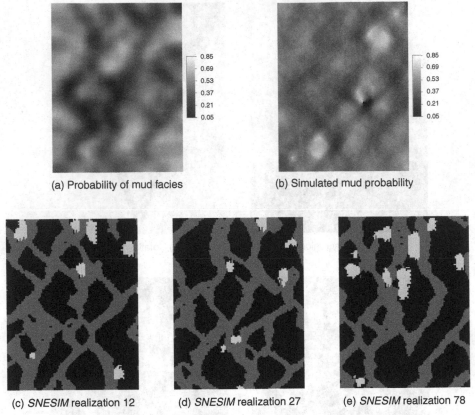

(a) Probability of mud facies

(b) Simulated mud probability

(c) *SNESIM* realization 12

(d) *SNESIM* realization 27

(e) *SNESIM* realization 78

Figure 8.28 [Example 4] Input target mud facies probability, simulated mud probability from 100 *SNESIM* realizations and three *SNESIM* realizations (black: mud facies; gray: sand channel; white: crevasse)

memory-demanding when the training image is large with a large number of categories and a large variety of different patterns. The mps algorithm *FILTER-SIM*, called filter-based simulation (Zhang *et al.*, 2006; Wu *et al.*, in press), has been proposed to circumvent these problems. The *FILTERSIM* algorithm is much less memory demanding yet with a reasonable CPU cost, and it can handle both categorical and continuous variables, but *FILTERSIM* has several shortcomings, see hereafter.

FILTERSIM utilizes a few linear filters to classify training patterns in a filter space of reduced dimension. Similar training patterns are stored in a class characterized by an average pattern called prototype. During simulation, the prototype closest to the conditioning data event is determined. A training pattern from that prototype class is then drawn, and pasted back onto the simulation grid. This is similar to constructing a jigsaw puzzle from stacks of similar pieces.

Instead of saving faithfully all training replicates in a search tree as does *SNESIM*, *FILTERSIM* only saves the central location of each training pattern in memory, hence reducing RAM demand.

The original *FILTERSIM* algorithm has been coded into two programs: *FIL-TERSIM_CONT* for continuous simulation and *FILTERSIM_CATE* for categorical simulation. The following generic descriptions apply to both programs.

Filters and scores

A filter is a set of weights associated with a specific data configuration/template of size J, $T_J = \{\mathbf{u}_0; \mathbf{h}_i, i = 1, \ldots, J\}$. Each node $\mathbf{u}_i$ of the template is defined by a relative offset vector $\mathbf{h}_i = (x, y, z)_i$ from the template center $\mathbf{u}_0$ and is associated with a specific filter value or weight f_i. The offset coordinates x, y, z are integer values. For a J-nodes template, its associated filter is $\{f(\mathbf{h}_i); i = 1, \ldots, J\}$. The filter configuration can be of any shape: Fig. 8.29a shows an irregular shaped filter and Fig. 8.29b gives a block-shaped filter of size $5 \times 3 \times 5$. Many different filters are considered and applied to each training pattern.

A search template is used to define patterns from a training image. The search template of *FILTERSIM* must be rectangular of size (n_x, n_y, n_z), where n_x, n_y, n_z

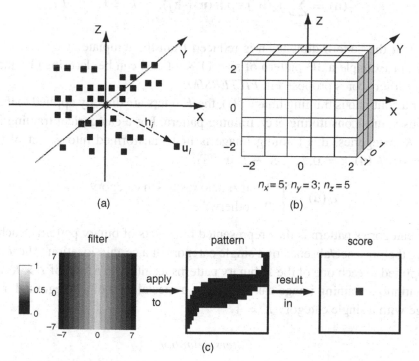

Figure 8.29 Filter and score: (a) a general template; (b) a cube-shaped template; (c) from filter to score value

are odd positive integers. Each node of this search template is recorded by its relative offset to the centroid. Figure 8.29b shows a search template of size $5 \times 3 \times 5$.

FILTERSIM requires that all filter configurations be the same as the search template, such that the filters can be applied to the training pattern centered at location **u**. Each training pattern is then summarized by a set of score values $S_T(\mathbf{u})$, one score per filter:

$$S_T(\mathbf{u}) = \sum_{j=1}^{J} f(\mathbf{h}_j) \cdot pat(\mathbf{u} + \mathbf{h}_j), \qquad (8.7)$$

where $pat(\mathbf{u} + \mathbf{h}_i)$ is the pattern nodal value, $J = n_x \times n_y \times n_z$. Figure 8.29c illustrates the process of creating a filter score value with a specific 2D filter.

Clearly, one filter is not enough to capture the information carried by any given training pattern. A set of F filters should be designed to capture the diverse characteristics of a training pattern. These F filters create a vector of F scores summarizing each training pattern, Eq. (8.7) is rewritten as:

$$S_T^k(\mathbf{u}) = \sum_{j=1}^{J} f_k(\mathbf{h}_j) \cdot pat(\mathbf{u} + \mathbf{h}_j), \qquad k = 1, \ldots, F. \qquad (8.8)$$

Note that the pattern dimension is reduced from the template size $n_x \times n_y \times n_z$ to F. For example a 3D pattern of size $11 \times 11 \times 3$ can be described by the nine default filter scores proposed in *FILTERSIM*.

For a continuous training image (Ti), the F filters are directly applied to the continuous values constituting each training pattern. For a categorical training image with K categories, this training image is first transformed into K set of binary indicators $I_k(\mathbf{u})$, $k = 0, \ldots, K - 1$, $\mathbf{u} \in$ Ti:

$$I_k(\mathbf{u}) = \begin{cases} 1 & \text{if } \mathbf{u} \text{ belongs to } k\text{th category} \\ 0 & \text{otherwise} \end{cases}. \qquad (8.9)$$

A K-categories pattern is thus represented by K sets of binary patterns, each indicating the presence/absence of a single category at a certain location. The F filters are applied to each one of the K binary patterns resulting in a total of $F \times K$ scores. A continuous training image can be seen as a special case of a categorical training image with a single category $K = 1$.

Filters definition

FILTERSIM accepts two filter definitions: the default filters defined hereafter and user-defined filters.

By default, *FILTERSIM* provides three filters (average, gradient and curvature) for each X/Y/Z direction, with the filters configuration being identical to that of the search template. Let n_i be the template size in the i direction (i denotes either X, Y or Z), $m_i = (n_i - 1)/2$, and $\alpha_i = -m_i, \ldots, +m_i$ be the filter node offset in the i direction, then the default filters are defined as:

- average filter: $f_1^i(\alpha_i) = 1 - \frac{|\alpha_i|}{m_i} \in [0, 1]$
- gradient filter: $f_2^i(\alpha_i) = \frac{\alpha_i}{m_i} \in [-1, 1]$
- curvature filter: $f_3^i(\alpha_i) = \frac{2|\alpha_i|}{m_i} - 1 \in [-1, 1]$

The default total is six filters for a 2D search template and nine in 3D.

Users can also design their own filters and enter them into a data file. The filter data file should follow the following format (see Fig. 8.30).

- The first line must be an integer number indicating the total number of filters included in this data file. Starting from the second line, list each filter definition one by one.
- For each filter, the first line gives the filter name which must be a string and the weight associated to the corresponding filter score (this weight is used later for pattern classification). In each of the following lines, list the offset (x, y, z) of each template node and its associated weight ($f(x, y, z)$). The four numbers must be separated by spaces.

Although the geometry of the user-defined filters can be of any shape and any size, only those filter nodes within the search template are actually retained for the score calculation to ensure that the filter geometry is the same as that of the

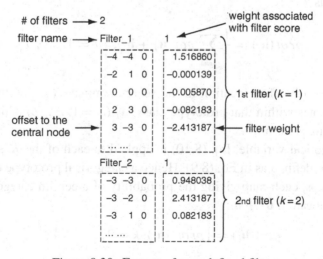

Figure 8.30   Format of user-defined filter

search template. For those nodes in the search template but not in the filter template, *FILTERSIM* adds dummy nodes associated with a zero filter value. There are many ways to create the filters, Principal Component Analysis (PCA) (Jolliffe, 1986) is one alternative.

Tip 6

For simple training geometries, the last curvature filter of the three default filters in each direction might be redundant. One may consider dropping this curvature filter in each direction using the user-defined filters, thus reducing the total number of filters to six in 3D; this would result in substantial CPU savings, especially for categorical simulations, at the risk of poorer pattern reproduction.

Pattern classification

Sliding the F filters over a K-category training image will result in $F \cdot K$ score maps, where each local training pattern is summarized by a $F \cdot K$-length vector in the filter score space. In general, $F \cdot K$ is much smaller than the size of the filter template T, hence the dimension reduction is significant.

Similar training patterns will have similar $F \cdot K$ scores. Hence by partitioning the filter score space, similar patterns can be grouped together. Each pattern class is represented by a pattern prototype *prot*, defined as the point-wise average of all training patterns falling into that class. A prototype has the same size as the filter template, and is used as the pattern group ID (identification number).

For a continuous training image, a prototype associated with search template T_J is calculated as:

$$prot(\mathbf{h}_i) = \frac{1}{c} \sum_{j=1}^{c} pat(\mathbf{u}_j + \mathbf{h}_i), \quad i = 1, \dots, J \tag{8.10}$$

where $\mathbf{h}_i$ is the ith offset location in the search template T_J, c is the number of training replicates within that prototype class; $\mathbf{u}_j$ ($i = 1, \dots, c$) is the center of a specific training pattern.

For a categorical variable, Eq. (8.10) is applied to each of the K sets of binary indicator maps defined as in Eq. (8.9). Hence a categorical prototype consists of K proportion maps, each map giving the probability of a certain category to prevail at a template location $\mathbf{u}_j + \mathbf{h}_i$:

$$\mathbf{prot}(\mathbf{h}_i) = \left\{ prot^k(\mathbf{h}_i), k = 1, \dots, K \right\}, \tag{8.11}$$

where $prot^k(\mathbf{h}_i) = P(z(\mathbf{u} + \mathbf{h}_i) = k)$.

For maximal CPU efficiency, a two-step partition approach is proposed.

1. Group all training patterns into some rough pattern clusters using a fast classification algorithm; these rough pattern clusters are called parent classes. Each
   parent class is characterized by its own prototype.
2. Partition those parent classes that have both too many and too diverse patterns in it using the same (previous) classification algorithm. The resulting
   sub-classes are called children classes. These children classes might be further partitioned if they contain too many and too diverse patterns. Each final
   child class is characterized by its own prototype of type Eq. (8.10).

For any class and corresponding prototype, the diversity is defined as the
averaged filter variance:

$$V = \frac{1}{F \cdot K} \sum_{k=1}^{F \cdot K} \omega_k \cdot \sigma_k^2 \qquad (8.12)$$

where $\omega_k \geq 0$ is the weight associated with the kth filter score, $\sum_{k=1}^{F \cdot K} \omega_k = 1$. For
the default filter definition, ω_k goes decreasing, 3, 2 and 1 for average, gradient
and curvature filters, respectively. For the user-defined filters, the ω_k value for each
filter must be specified in the filter data file (Fig. 8.30):

$\sigma_k^2 = \frac{1}{c} \sum_{i=1}^c (S_k^i - m_k)^2$ is the variance of the kth score value over the c
  replicates;

S_k^i is the score of the ith replicate of kth filter score defining the prototype;

$m_k = \frac{1}{c} \sum_{i=1}^c S_{k,l}^i$ is the mean value of the kth score value over the c
  replicates.

The prototypes with diversity higher than a threshold (calculated automatically)
and with too many replicates are further partitioned.

This proposed two-step partition approach allows finding quickly the prototype
which is the closest to the data event. Consider a case with 3000 prototypes (parents and children), without the two-step partition it would take at each node 3000
distance comparisons; with a two-step partition considering 50 parent prototypes
it would take 50 comparisons to find the best parent prototype, then in average 60
comparisons to find the best child prototype, thus in average a total of 110 distance
comparisons.

Partition method

Two classification methods are provided: cross partition (Zhang *et al.*, 2006) and
K-Mean clustering partition (Hartigan, 1975). The cross partition consists of partitioning independently each individual filter score into equal frequency bins (see

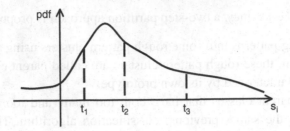

(a) Divide scores into 4 equal frequency bins

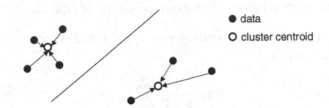

(b) Group scores into 2 clusters with K-Mean clustering

Figure 8.31 Two classification methods

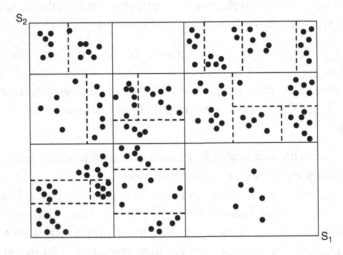

Figure 8.32   Illustration of cross partition in a 2-filter score space. Each dot represents a local training pattern; the solid lines show first parent partition ($M = 3$); the dash lines give the secondary children partition ($M = 2$)

Fig. 8.31a). Given a score space of dimension $F \cdot K$, if each filter score is partitioned into M bins ($2 \le M \le 10$), thus the total number of parent classes is $M^{F \cdot K}$. However, because the filter scores are partitioned independently of one another, many of these classes will contain no training patterns. Figure 8.32 shows the results

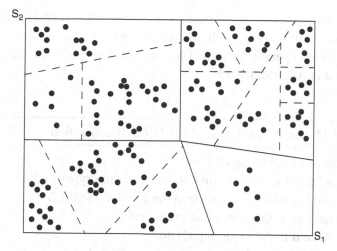

Figure 8.33 Illustration of K-Mean clustering partition in a 2-filter score space. Each dot represents a local training pattern; the solid lines show first parent partition ($M = 4$); the dash lines give the secondary children partition ($M = 3$)

of cross partition in a 2-filter score space using the proposed two-step approach splitting parent classes into children.

The cross partition approach is fast; however, it is rough and may lead to many classes having few or no replicates.

A much better but costlier partition method using K-Mean clustering is also proposed: given an input number of clusters, the algorithm will find the optimal centroid of each cluster, and assign each training pattern to a specific cluster according to a distance between the training pattern and the cluster centroid (see Fig. 8.31b). This K-Mean clustering partition is one of the simplest unsupervised learning algorithms; it creates better pattern groups all with a reasonable number of replicates, however it is slow compared to the cross partition. Also the number of clusters is critical to both CPU cost and the final simulation results. Figure 8.33 shows the results of K-Mean clustering partition in a 2-filter score space with the proposed two-step approach.

Single grid simulation

After creating the prototype list (for all parents and children) built from all the training patterns, one can proceed to generate simulated realizations.

The classic sequential simulation paradigm defined in Section 3.3.1 is extended to pattern simulation. At each node **u** along the random path visiting the simulation grid G, a search template T of the same size as the filter template is used to extract the conditioning data event $dev(\mathbf{u})$. The prototype closest to that data event, based on some distance function, is found. Next a pattern *pat* is randomly drawn from that closest prototype class, and is pasted onto the simulation grid G. The inner part of the pasted pattern is frozen as hard data, and will not be revisited during simulation on the current (multiple) grid. The simple single grid *FILTERSIM* approach is summarized in Algorithm 8.15.

Algorithm 8.15   Simple, single grid *FILTERSIM* simulation

1: Create score maps with given filters
2: Partition all training patterns into classes and prototypes in the score space
3: Relocate hard conditioning data into the simulation grid G
4: Define a random path on the simulation grid G
5: **for** Each node **u** in the random path **do**
6:     Extract the conditioning data event dev centered at **u**
7:     Find the parent prototype $prot_p$ closest to dev
8:     **if** $prot_p$ has children prototype lists **then**
9:        Find the child prototype $prot_c$ closest to dev
10:       Randomly draw a pattern pat from $prot_c$
11:     **else**
12:       Randomly draw a pattern pat from $prot_p$
13:     **end if**
14:     Paste pat to the realization being simulated, and freeze the nodes within a central patch
15: **end for**

Distance definition

A distance function is used to find the prototype closest to a given data event dev. The distance between dev and any prototype is defined as:

$$d = \sum_{i=1}^{J} \omega_i \cdot |dev(\mathbf{u} + \mathbf{h}_i) - prot(\mathbf{u}_0 + \mathbf{h}_i)| \qquad (8.13)$$

where

     J is the total number of nodes in the search template T;
     ω_i is the weight associated to each template node;
     u is the center node of the data event;
     $\mathbf{h}_i$ is the node offset in the search template T;
     $\mathbf{u}_0$ is the center node location of the prototype.

Given three different data types: original hard data ($d = 1$), previously simulated values frozen as hard data ($d = 2$), other values informed by pattern pasting ($d = 3$). The above weight ω_i is defined as

$$
\omega_i = \begin{cases} W_1/N_1 & : \text{ hard data } (d=1) \\ W_2/N_2 & : \text{ patch data } (d=2) \text{ ,} \\ W_3/N_3 & : \text{ other } (d=3) \end{cases}
$$

where W_d ($d = 1, 2, 3$) is the weight associated with data type d, and N_d is the number of nodes of data type d within the data event dev. It is required that $W_1 + W_2 + W_3 = 1$, and $W_1 \geq W_2 \geq W_3$, to emphasize the impact of hard data and data frozen as hard (inner patch values). The weights W are user input parameters, and the default values are 0.5, 0.3 and 0.2 for hard, patch, and other data type respectively.

Note that the template of a data event dev is usually not fully informed, thus only those informed nodes are retained for the distance calculation.

Multiple grid simulation

Similar to the *SNESIM* algorithm, the multiple grid simulation concept (see Section 8.2.1) is used to capture the large scale structures of the training image with a large but coarse template T. In the gth ($1 \leq g \leq L$) coarse grid, the filters defined on the rescaled template T^g are used to calculate the pattern scores. Sequential simulation proceeds from the coarsest grid to the finest grid. All nodes simulated in the coarser grid are re-simulated in the next finer grid.

The template is expanded isotropically as described in the *SNESIM* algorithm (Section 8.2.1). The *FILTERSIM* multiple grid simulation is summarized in Algorithm 8.16.

Algorithm 8.16 *FILTERSIM* simulation with multiple grids

1: **repeat**
2:     For the gth coarse grid, rescale the geometry of the search template, the inner patch template and the filter template
3:     Create score maps with the rescaled filters
4:     Partition the training patterns into classes and corresponding prototypes
5:     Define a random path on the coarse simulation grid G^g
6:     Relocate hard conditioning data into the current coarse grid G^g
7:     Perform simulation on current grid G^g (Algorithm 8.15)
8:     If $g \neq 1$, delocate hard conditioning data from the current coarse grid G^g
9: **until** All multi-grids have been simulated

Score-based distance calculation

The distance function defined in Eq. (8.13) works well for both continuous and categorical simulations over 2D grids. However, it becomes slow for 3D simulations, especially for multiple categorical variables. Given N_p prototype classes, and $J = n_x \times n_y \times n_z$ number of nodes in the search template (see Fig. 8.29), the total number of distance calculation time will be $\propto N_p \cdot J$. For a large 3D training image, N_p might be in the order of 10^4, hence the $N_p \cdot J$ distance calculations using Eq. (8.13) will be very time consuming.

One solution is to reduce the dimensions of both the data event *dev* and the training prototypes *prot* from J to $F \cdot K$ using the pre-defined filters, where F is the number of filters and K is the number of categories ($K = 1$ for a continuous variable). Then the distance function Eq. (8.13) can be modified as:

$$d = \sum_{i=1}^{F \cdot K} \left| S_{dev}^i(\mathbf{u}) - S_{prot}^i(\mathbf{u}_0) \right|, \tag{8.14}$$

where S is the score value. Because $F \cdot K << J$, the distance calculation with Eq. (8.14) is much faster than with Eq. (8.13). Hence the overall *FILTERSIM* simulation will be speeded up significantly.

However, in order to apply filters to the data event *dev*, that *dev* must be fully informed. With the multiple grid simulation concept, any coarse grid ($g > 1$) is only partially informed. The challenge here is how to fill in those uninformed locations. Because in 3D simulation, more than 85% of time is spent on the finest grid, the dual template concept (Arpat, 2004) is used to fill in the uninformed locations on the penultimate grid ($g = 2$). A dual template has the same spatial extent as the expanded next coarser grid template, but with all the fine grid nodes falling into that template present and informed. For instance, in Fig. 8.20 the expanded coarse grid template has the same number (9) of nodes as the finest grid template; the dual template has the same extent as the coarse grid template but with all the 25 nodes inside it all informed. Boundary nodes received a special treatment.

The *FILTERSIM* simulation in the last (finest) grid is described as in Algorithm 8.17.

Soft data integration

The *FILTERSIM* algorithm allows users to constrain simulations to soft data defined over the same simulation grid. The soft data when simulating a continuous variable should be a spatial trend (local varying mean) of the attribute being modeled, hence only one soft data set is allowed with the unit of the variable being simulated. For categorical training images, there is one soft data set per category.

Algorithm 8.17 Fine grid *FILTERSIM* simulation with score-based distance

---

1: After completing the simulation on the penultimate coarse grid, use the dual template to fill in the uninformed locations on the finest grid
2: Create score maps with the pre-defined filters
3: Partition the training patterns into classes and corresponding prototypes
4: Apply filters to all pattern prototypes
5: Define a random path on the finest simulation grid G^1
6: Relocate hard conditioning data into the finest grid G^1
7: **for** Each node $\mathbf{u}$ in the random path **do**
8:    Extract the conditioning data event dev centered at $\mathbf{u}$
9:    Calculate dev scores S_{dev} with the same pre-defined filters
10:    Find the best prototype $prot$ closest to dev using Eq. (8.14)
11:    Randomly draw a pattern pat from $prot$
12:    Paste pat to the realization being simulated, and freeze the nodes within a central patch
13: **end for**

Each soft cube is a probability field related to the presence/absence of a category at each simulation grid node $\mathbf{u}$, hence there is a total of K probability cubes.

- The procedure of integrating soft data for a continuous variable is described in Algorithm 8.18. The soft data event $sdev$ is used to fill in the data event dev: at any uninformed location $\mathbf{u}_j$ of the data event template, set its value to the soft data value at the same location ($dev(\mathbf{u}_j) = sdev(\mathbf{u}_j)$). Because these soft data contribute to the prototype selection, the choice of the sampled pattern is constrained by the local trend.
- For a categorical variable, the original training image has been internally transformed into K binary indicator maps (category probabilities) through Eq. (8.9), thus each resulting prototype is a set of K probability templates (Eq. (8.11)). At each simulation location $\mathbf{u}$, the prototype closest to the data event dev is a probability vector $\mathbf{prob}(\mathbf{u})$. The same search template T is used to retrieve the soft data event $sdev(\mathbf{u})$ at the location $\mathbf{u}$. The tau model (see Section 3.10), is used to combine $sdev(\mathbf{u})$ and $\mathbf{prob}(\mathbf{u})$ pixel-wise at each node $\mathbf{u}_j$ of the search template T into a new probability cube dev^*. A prototype is found which is closest to dev^*, and a pattern is randomly drawn and pasted onto the simulation grid. The detailed procedure of integrating soft probability data for categorical attributes is presented in Algorithm 8.19.

Algorithm 8.18   Data integration for continuous variables

1: At each node **u** along the random path, use the search template T to extract both the data event dev from the realization being simulated, and the soft data event $sdev$ from the soft data field
2: **if** dev is empty (no informed data) **then**
3:     Replace dev by $sdev$
4: **else**
5:     Use $sdev$ to fill in dev at all uninformed data locations within the search template T centered at **u**
6: **end if**
7: Use dev to find the closest prototype, and proceed to simulation of the node **u**

Algorithm 8.19   Data integration for categorical variables

1: At each node **u** along the random path, use the search template T to retrieve both the data event dev from the realization being simulated and the soft data event $sdev$ from the input soft data field
2: **if** dev is empty (no informed data) **then**
3:     Replace dev by $sdev$, and use the new dev to find the closest prototype.
4: **else**
5:     Use dev to find the closest prototype **prot**
6:     Use tau model to combine prototype **prot** and the soft data event $sdev$ into a new data event dev^* as the local probability map
7:     Find the prototype closest to dev^*, and proceed to simulation
8: **end if**

> **Tip 7**
> *The present algorithms for integrating soft data in* FILTERSIM *are stopgaps and should be rapidly replaced by better ones where the soft data would be used for more than the distance calculations. If the user is very confident with the soft probability cubes (for the categorical simulations), then increase the τ_2 value to 10 or 15 which results in increasing the soft data impact.*

Accounting for local non-stationarity

The same region concept as presented in the *SNESIM* algorithm (Section 8.2.1) is introduced here to account for local non-stationarity. It is possible to perform *FILTERSIM* simulation over regions, with each region associated with a specific training image and its own parameter settings. See Section 8.2.1 of the *SNESIM* algorithm for greater details.

The rotation and affinity concept is not coded in the current *FILTERSIM* algorithm; however, it can be done with the region concept by rotating or rescaling the training image explicitly. The Python script, introduced in Section 10.2.2, can be used to automate this task.

Parameters description: **FILTERSIM_CONT**

The *FILTERSIM_CONT* for continuous variable simulations can be invoked from *Simulation* → *filtersim_cont* in the upper part of the Algorithm Panel. Its interface has four pages: "General", "Conditioning", "Region" and "Advanced" (see Fig. 8.34). The *FILTERSIM_CONT* parameters is presented page by page in the following. The text inside "[ ]" is the corresponding keyword in the *FILTERSIM_CONT* parameter file.

1. **Simulation Grid Name** [GridSelector_Sim] The name of grid on which simulation is to be performed.
2. **Property Name Prefix** [Property_Name_Sim] The name of the property to be simulated.
3. **# of Realizations** [Nb_Realizations] Number of realizations to be simulated.
4. **Seed** [Seed] A large odd integer to initialize the pseudo-random number generator.
5. **Training Image | Object** [PropertySelector_Training.grid] The name of the grid containing the training image.
6. **Training Image | Property** [PropertySelector_Training.property] The training image property, which must be a continuous variable.
7. **Search Template Dimension** [Scan_Template] The size of the 3D template used to define the filters. The same template is used to retrieve training patterns and data events during simulation.
8. **Inner Patch Dimension** [Patch_Template_ADVANCED] The size of the 3D patch of simulated nodal values frozen as hard data during simulation.
9. **Match Training Image Histogram** [Trans_Result] This flag indicates whether the simulated result on the penultimate coarse grid should be transformed to honor the training image statistics. If marked, perform internal *TRANS* (Section 9.1) on those simulated results. The default is not to use internal *TRANS*.
10. **Hard Data | Object** [Hard_Data.grid] The grid containing the hard conditioning data. The hard data object must be a point-set. The default input is "None", which means no hard conditioning data are used.

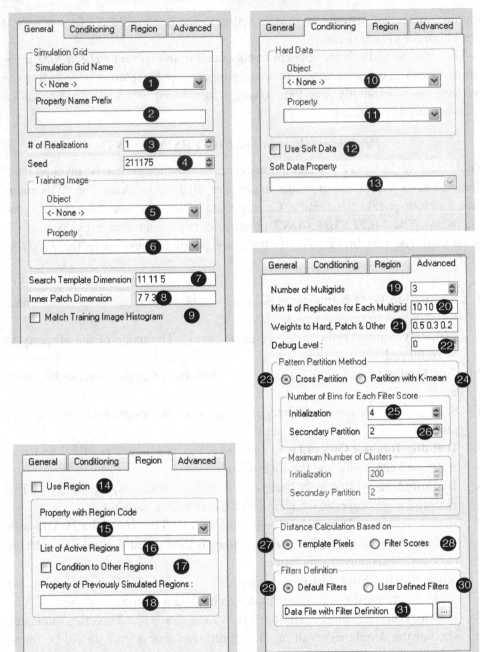

Figure 8.34 *FILTERSIM_CONT* interface

11. **Hard Data | Property** [Hard_Data.property]   The property of the hard conditioning data, which must be a continuous variable.

12. **Use Soft Data** [Use_SoftField]   This flag indicates whether the simulation should be conditioned to prior local probability cubes. If marked, perform *FILTERSIM* conditional to soft data. The default is not to use soft data.

13. **Soft Data Property** [SoftData]   Selection for local soft data conditioning. Note that only one soft conditioning property is allowed which is treated as a local varying mean, and the soft data must be given over the same simulation grid as defined in item 1.

14. **Use Region** [Use_Region]   The flag indicates whether to use region concept. If marked (set as 1), perform *SNESIM* simulation with the region concept; otherwise perform *FILTERSIM* simulation over the whole grid.

15. **Property with Region Code** [Region_Indicator_Prop]   The property containing the index coding of the regions, must be given over the same simulation grid as defined in (1). The region code ranges from 0 to $N_R - 1$ where N_R is the total number of regions.

16. **List of Active Regions** [Active_Region_Code]   Input the index of region to be simulated, or indices of the regions to be simulated simultaneously. If simulation with multiple regions, the input region indices (codes) should be separated by spaces.

17. **Condition to Other Regions** [Use_Previous_Simulation]   The option to perform region simulation conditional to data from other regions.

18. **Property of Previously Simulated Regions** [Previous_Simulation_Pro] The property simulated in the other regions. The property can be different from one region to another. See Section 8.2.1.

19. **# of Multigrids** [Nb_Multigrids_ADVANCED]   The number of multiple grids to consider in the multiple grid simulation. The default value is 3.

20. **Min # of Replicates for Each Grid** [Cmin_Replicates]   A pattern prototype split criteria. Only those prototypes with more than Cmin_Replicates can be further divided. Input a Cmin_Replicates value for each multiple coarse grid. The default value is 10 for each multigrid.

21. **Weights to Hard, Patch & Other** [Data_Weights]   The weights assigned to different data types (hard data, patched data and all the other data). The sum of these weights must be 1. The default values are 0.5, 0.3 and 0.2.

22. **Debug Level** [Debug_Level]   The flag controls the output in the simulation grid. The larger the debug level, the more outputs from *FILTERSIM* simulation:

    - if 0, then only the final simulation result is output (default value);
    - if 1, then the filter score maps associated with the original fine grid search template are also output in the training image grid;

- if 2, then the simulation results on the coarse grids are output in addition to the outputs of options 0 and 1;
- if 3, then output one more property in the training image grid, which is the indicator map for all parent prototypes.

23. **Cross Partition** [CrossPartition] Perform pattern classification with cross partition method (default option).

24. **Partition with K-Mean** [KMeanPartition] Perform pattern classification with K-Mean clustering method. Note that 'Cross Partition' and 'Partition with K-mean' are mutually exclusive.

25. **Number of Bins for Each Filter Score | Initialization** [Nb_Bins_ADVANCED] The number of bins for parent partition when using cross partition; the default value is 4. Or the number of clusters for parent partition when using K-Mean partition; the default value is 200.

26. **Number of Bins for Each Filter Score | Secondary Partition** [Nb_Bins_ADVANCED2] The number of bins for children partition when using cross partition; the default value is 2. Or the number of clusters for children partition when using K-Mean partition; the default value is 2.

27. **Distance Calculation Based on | Template Pixels** [Use_Normal_Dist] The distance as the pixel-wise sum of differences between the data event values and the corresponding prototype values (Eq. (8.13)). This is the default option.

28. **Distance Calculation Based on | Filter Scores** [Use_Score_Dist] The distance is defined as the sum of differences between the data event scores and the pattern prototype scores.

29. **Default Filters** [Filter_Default] The option to use the default filters provided by *FILTERSIM*: 6 filters for a 2D search template and 9 filters for a 3D search template.

30. **User Defined Filters** [Filter_User_Define] The option to use user's own filter definitions. Note that 'Default' and 'User Defined' are mutually exclusive.

31. **The Data File with Filter Definition** [User_Def_Filter_File] Input a data file with the filter definitions (see Fig. 8.30). This parameter is ignored if Filter_User_Define is set to 0.

Parameters description: **FILTERSIM_CATE**

The *FILTERSIM_CATE* algorithm for categorical variable simulations can be invoked from *Simulation* → *filtersim_cate* in the upper part of the Algorithm Panel. Its interface has four pages: "General", "Conditioning", "Region" and "Advanced" (the first two pages are given in Fig. 8.35, and the last two pages are given in Fig. 8.34). Because most of *FILTERSIM_CATE* parameters are similar to those of

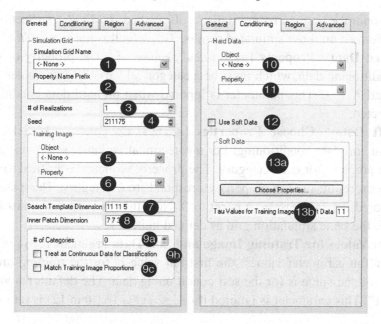

Figure 8.35 *FILTERSIM_CATE* interface

FILTERSIM_CONT program, only the parameters unique to *FILTERSIM_CATE* (items 6, 9, 11 and 13) are presented in the following. Refer to *FILTERSIM_CONT* parameter descriptions for all other parameters. The text inside "[ ]" is the corresponding keyword in the *FILTERSIM_CONT* parameter file.

6. Training Image | Property [PropertySelector_Training.property] The training image property, which must be a categorical variable whose value must be between 0 and $K - 1$, where K is the number of categories.

9a. # of Categories [Nb_Facies] The total number of categories when working with categorical variable. This number must be consistent with the number of categories in the training image.

9b. Treat as Continuous Data for Classification [Treat_Cate_As_Cont] The flag to treat the categorical training image as a continuous training image for pattern classification (but the simulation is still performed with categorical variables). With this option, the F filters are directly applied on the training image without having to transform the categorical variable numerical code into K sets of binary indicators, hence the resulting score space is of dimension F instead of $F \cdot K$, and it is faster. Note that the training image specific categorical coding will affect the simulation results.

9c. Match Training Image Proportions [Trans_Result] This flag indicates whether the simulated result on the next to last grid should be transformed to honor the training image statistics. If marked, *TRANSCAT* (Section 9.2) is performed on those simulated results. The default is not to use *TRANSCAT*.

11. **Hard Data | Property** [Hard_Data.property]   The property of the hard conditioning data, which must be a categorical variable with values between 0 and $K - 1$. This parameter is ignored when no hard data conditioning is selected.

13a. **Soft Data | Choose Properties** [SoftData_properties]   Selection for local soft data conditioning. For a categorical variable, select one and only one property for each category. The property sequence is critical to the simulation result: the kth property corresponds to k category. This parameter is ignored if Use_ProbField is set to 0. Note that the soft data must be given over the same simulation grid as defined in item 1.

13b. **Tau Values for Training Image and Soft Data** [TauModelObject]   Input **two** Tau parameter values: the first tau value is for the training image, the second tau value is for the soft conditioning data. The default tau values are "1 1". This parameter is ignored if Use_SoftField (item 12) is set to 0.

Examples

In this section, *FILTERSIM* is run for both unconditional and conditional simulations. The first example illustrates the ability of *FILTERSIM* algorithm to handle continuous variables with *FILTERSIM_CONT*; the other three examples demonstrate *FILTERSIM* algorithm with categorical training images with *FILTERSIM_CATE*.

EXAMPLE 1: 3D simulation of continuous seismic data

The *FILTERSIM* algorithm is used to complete a 3D seismic image. Figure 8.36a shows a 3D seismic image with a large center area uninformed because of shadow effect. The whole grid is of size $450 \times 249 \times 50$, and the percentage of the missing data is 24.3%. The goal here is to fill in those missing data locations by extending the geological information available in the neighboring areas. The North part of the original seismic image is retained as the training image, which is of size $150 \times 249 \times 50$, see the area in the rectangular box at the left of the white boundary.

For the *FILTERSIM* simulation, the size of search template is $21 \times 21 \times 7$, the size of patch template is $15 \times 15 \times 5$, the number of multiple grids is 3, and the number of bins for parent cross partition is 3. All the known data are used for hard conditioning. One *FILTERSIM* realization is given in Fig. 8.36b. The simulation area lies between the white line and the black line. On the simulation the layering structures are extended from the conditioning area to the simulation area, and the horizontal large scale structures are

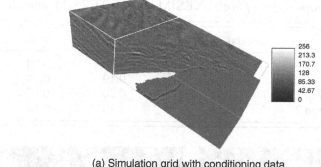

(a) Simulation grid with conditioning data

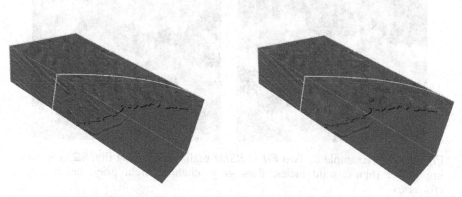

(b) Uninformed area filled in with *FILTERSIM*   (c) Uninformed area filled in with *SGSIM*

Figure 8.36 [Example 1] Use *FILTERSIM* and *SGSIM* to fill in the uninformed area of a 3D seismic cube

reasonably well reproduced. For comparison, the variogram-based algorithm *SGSIM* is also used to fill in the empty area with the variogram modeled from the training image. Figure 8.36c shows one *SGSIM* realization, in which the layering structures are completely lost.

EXAMPLE 2: 2D unconditional simulation

This example relates to a four facies unconditional simulation using Fig. 8.25a as training image. The search template is of size $23 \times 23 \times 1$, and the patch template is of size $15 \times 15 \times 1$. The number of multiple grid is 3, and the minimum number of replicates for each multiple grid is 10. Cross partition is used with 4 bins for the parent classification and 2 bins for the children classification. Figure 8.37 shows two *FILTERSIM* realizations, which depict a good training pattern reproduction. The facies proportions for these two realizations are given in Table 8.2. Compared to the *SNESIM* simulation of Fig. 8.25b, the *FILTERSIM* algorithm appears to better capture the large scale channel structures.

Table 8.2 *Facies proportions of both* SNESIM *and* FILTERSIM *simulations*

	mud background	sand channel	levee	crevasse
training image	0.45	0.20	0.20	0.15
SNESIM realization	0.44	0.19	0.20	0.17
FILTERSIM realization 1	0.51	0.20	0.17	0.12
FILTERSIM realization 2	0.53	0.18	0.18	0.11

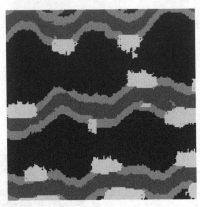

Figure 8.37 [Example 2] Two *FILTERSIM* realizations using Fig. 8.25a as training image (black: mud facies; dark gray: channel; light gray: levee; white: crevasse)

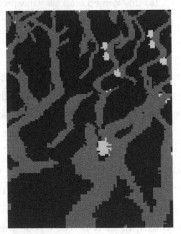

(a) Simulation with affinity regions   (b) Simulation with rotation regions

Figure 8.38 [Example 3] *FILTERSIM* simulation with affinity regions and with rotation regions using Fig. 8.27c as training image (black: mud facies; gray: sand channel; white: crevasse)

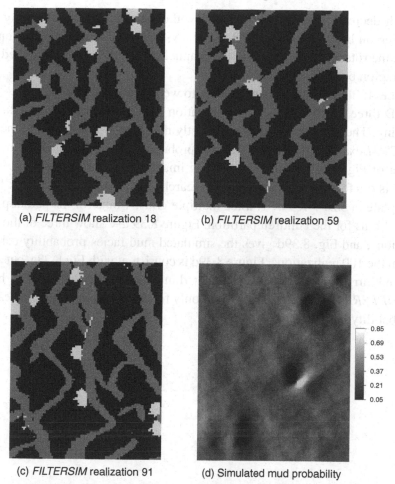

(a) *FILTERSIM* realization 18   (b) *FILTERSIM* realization 59

(c) *FILTERSIM* realization 91   (d) Simulated mud probability

Figure 8.39 [Example 4] Three *FILTERSIM* realizations (black: mud facies; gray: sand channel; white: crevasse) and the simulated mud probability obtained from 100 *FILTERSIM* realizations. The training image is given in Fig. 8.27c

EXAMPLE 3: 2D unconditional simulation with affinity and rotation

In this example, the region concept is used to account for local non-stationarity. The simulation field is of size $100 \times 130 \times 1$, the same as used in the third example of *SNESIM* algorithm (see Fig. 8.27a). The 2D training image is given in Fig. 8.27c. The affinity regions were given in Fig 8.27a, the rotation regions in Fig 8.27b. The region settings are exactly the same as used for the third *SNESIM* example.

FILTERSIM is performed with a search template of size $11 \times 11 \times 1$, patch template of size $7 \times 7 \times 1$, three multiple grids, 4 bins for the parent partition and 2 bins for the children partition. Figure 8.38a shows one *FILTERSIM* simulation using the affinity region only; it reflects the channel width decrease from South to North without significant discontinuity across the region boundaries. Figure 8.38b shows one *FILTERSIM* realization using only the rotation region. Again, the channel continuity is well preserved across the region boundaries.

EXAMPLE 4: 2D simulation conditioning to well data and soft data

A 2D three facies *FILTERSIM* simulation is performed with soft data conditioning. The problem settings are exactly the same as those used in the fourth *SNESIM* example (see Fig. 8.28): the probability fields are taken from the last layer of Fig. 8.26 c–e; and the training image is given in Fig. 8.27c. *FILTERSIM* is run for 100 realizations with a search template of size $11 \times 11 \times 1$, patch template of size $7 \times 7 \times 1$, three multiple grids, 4 bins for the parent partition and 2 bins for the children partition. Figure 8.39 a–c show three of those realizations, and Fig. 8.39d gives the simulated mud facies probability calculated from the 100 realizations. Figure 8.39d is consistent with Fig.8.28a, but slightly more blurry than the *SNESIM* simulated mud probability (Fig.8.28b), because in *FILTERSIM* the soft data are used only for distance calculation instead of as probability field directly.

9

Utilities

This chapter presents service algorithms helpful in many geostatistical studies. The first algorithm is the histogram transformation *TRANS*. Program *TRANS* in Section 9.1 allows the transforming of any histogram into any other histogram by matching their quantiles. The second algorithm (Section 9.2) is a proportion transformation appropriate for categorical variables. Program *TRANSCAT* not only matches target proportions but also preserves the image structures. The third algorithm is *POSTKRIGING* (Section 9.3) which extracts useful information from kriging or indicator kriging maps. Program *POSTSIM* (Section 9.4) performs the same tasks but on a set of stochastic realizations resulting from any of the simulation algorithms presented in Chapter 8. Section 9.5 presents algorithm *NU-TAU MODEL* related to the nu/tau model, see Section 3.10: this algorithm allows the combining of different probabilities stored as properties. Section 9.6 presents utility program *BCOVAR*, which allows the user to calculate the covariance map between any point or block, and the covariance value between any points or blocks. Section 9.7 presents program *IMAGE PROCESSING*, used to perform scaling and rotation on a Cartesian grid. These operations are particularly useful for preparing a training image. *MOVING WINDOW* in Section 9.8 calculates local statistics such as moving average, moving variance, the default *FILTERSIM* filters, a Gaussian low pass filter and the Sobel filter for edge detection; it also accepts user-defined filters. Finally, a training image generator *TIGENERATOR* is presented in Section 9.9.

9.1 *TRANS*: histogram transformation

Algorithm *TRANS* allows the user to transform any histogram of a continuous attribute into any other one. For example, the Gaussian simulation algorithms (*SGSIM* and *COSGSIM*), as described in Sections 8.1.2 and 8.1.3, apply only to Gaussian variables. If the attribute of interest does not display a Gaussian histogram, it is possible to transform that attribute into a Gaussian distribution, then work on that transformed variable. The histogram transformation is detailed in Algorithm 9.1.

Algorithm 9.1   Histogram transformation

1: **for** Each value z_k to be rank-transformed **do**
2:     Get the quantile from the source histogram associated with z_k, $p_k = F_Z(z_k)$
3:     Get the values y_k from the target distribution associated with p_k, $y_k = F_Y^{-1}(p_k)$
4:     **if** weighted transformation **then**
5:         Applied weighting to the transform
6:     **end if**
7: **end for**

Algorithm *TRANS* transforms a property following a source distribution into a new variable that follows a target distribution. The transformation of a variable Z with a source cdf F_Z into variable Y with target cdf F_Y is written (Deutsch and Journel, 1998):

$$Y = F_Y^{-1}\left(F_Z\left(Z\right)\right). \tag{9.1}$$

When the data used to build the source distribution contain identical values, it is likely that the target distribution will not be matched. In that case, use the "break ties" option available when building the non-parametric distribution function, see Section 6.8.

Note: the transformation of a source distribution to a Gaussian distribution does not ensure that the new variable Y is multivariate Gaussian, it is only univariate Gaussian. One should check that the multivariate (or at least bivariate) Gaussian hypothesis holds for Y before performing Gaussian simulation. If the hypothesis is not appropriate, other algorithms that do not require Gaussianity, e.g. direct sequential simulation (*DSSIM*) or sequential indicator simulation (*SISIM*), should be considered.

Histogram transformation with conditioning data

It is possible to apply a weighting factor to control how much each specific original value should be transformed (Deutsch and Journel, 1998, p.228):

$$y = z - \omega(z - F_Y^{-1}(F_Z(z))). \tag{9.2}$$

When $\omega = 0$ then $y = z$, there is no transformation. When $\omega = 1$ then $y = F_Y^{-1}(F_Z(z)))$ which is the standard rank transform, see Eq. (9.1). The weight ω

can be set equal to a standardized kriging variance as provided by simple kriging with a unit sill variogram. At Z-data locations the kriging variance is zero, hence there is no transform and the datum value is unchanged: $y = z$. Away from data locations the kriging variance increases allowing for a larger transform. That option is to be used for slight adjustments of the marginal distribution. When the weights ω are used the transform is not rank-preserving anymore and the target distribution would only be matched approximately.

The histogram transformation algorithm is given in Algorithm 9.1.

Parameters description

The *TRANS* algorithm is activated from *Utilities* → *trans* in the Algorithm Panel. The *TRANS* interface contains three pages: "Data", "Source" and "Target" (see Fig. 9.1). The text inside "[ ]" is the corresponding keyword in the *TRANS* parameter file.

1. **Object Name** [grid]   Selection of the grid containing the properties to be transformed.
2. **Properties** [props]   Properties to be transformed.
3. **Suffix for output** [out_suffix]   The name for each output property consists of the original name plus the suffix entered here.
4. **Local Conditioning** [is_cond]   Enables the use of weights for the histogram transformation. This allows a gradual transformation away from the data.

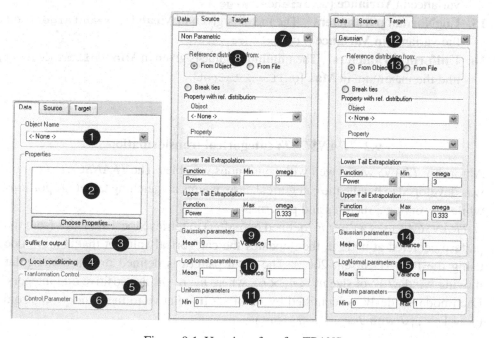

Figure 9.1 User interface for *TRANS*

5. **Weight Property** [cond_prop]   Property with weights for transformation of the histogram conditional to data, see Eq. (9.2). The standardized kriging variance is a good weighting option. Only required if **Local Conditioning** [is_cond] is selected.

6. **Control Parameter** [weight_factor]   Value between 0 and 1 adjusting the weights. Only required if **Local Conditioning** [is_cond] is selected.

7. **Source histogram** [ref_type_source]   Define the type of the source histogram: non-parametric, Gaussian, LogNormal, or Uniform (see items **8** to **11**).

8. **Non Parametric**   The non-parametric distribution is entered in [nonParamCdf_source], see Section 6.8 for format.

9. **Gaussian parameters**   The mean is given in **Mean** [G_mean_source] and the variance in **Variance** [G_variance_source].

10. **LogNormal parameters**   The mean is given in **Mean** [LN_mean_source] and the variance in **Variance** [LN_variance_source].

11. **Uniform parameters**   The minimum value is given in **Min** [Unif_min_source] and the maximum in **Max** [Unif_max_source].

12. **Target histogram** [ref_type_target]   Define the type of the target histogram: non-parametric, Gaussian, LogNormal, or Uniform (see items **13** to **16**).

13. **Non Parametric**   The non-parametric distribution is entered in [nonParamCdf_target], see Section 6.8 for format.

14. **Gaussian parameters**   The mean is given in **Mean** [G_mean_target] and the variance in **Variance** [G_variance_target].

15. **LogNormal parameters**   The mean is given in **Mean** [LN_mean_target] and the variance in **Variance** [LN_variance_target].

16. **Uniform parameters**   The minimum value is given in **Min** [Unif_min_target] and the maximum in **Max** [Unif_max_target].

9.2 *TRANSCAT*: categorical transformation

Algorithm *TRANS* works well with continuous variables. The *TRANSCAT* program was developed to handle the categorical case: it aims at matching target proportions while preserving structures and patterns.

TRANSCAT uses a filter F (defined hereafter) to retrieve local statistics from the pattern centered at node **u**, then updates the categorical value at that center node using the filter (Deutsch, 2002). The local pattern is defined over a rectangular moving window W of size $n_x \times n_y \times n_z$, where n_x, n_y, n_z are positive odd integers. Each node inside W is identified by its offset $\mathbf{h}_j$ relative to the window centroid ($j = 1, \ldots, J$ with $J = n_x \times n_y \times n_z$).

Similar to the *FILTERSIM* algorithm (Section 8.2.2), *TRANSCAT* accepts two alternative filter definitions: the default filter or any user-defined filter.

The default filter has the same geometry $\{\mathbf{h}_j : j = 1, \ldots, J\}$ as the moving window W, and the filter weight $f_{\mathbf{h}_j}$ is defined as follows:

1. the filter value of the central node is $V_c = \max\{n_x, n_y, n_z\}$;
2. all outer aureole nodes have unit filter value;
3. the filter values for the nodes along the major axes (X, Y and Z) are linearly interpolated as an integer between 1 and V_c based on their absolute offsets $|\mathbf{h}_j|$;
4. the filter at any other node is set equal to $V_c/(1+dist)$, where $dist$ is the square root distance between that node and the central node. The values are rounded to the nearest integer.

For example, a default filter of size $5 \times 5 \times 5$ is as follows.

```
Layers 1,5    Layers 2,4    Layer 3
1 1 1 1 1     1 1 1 1 1     1 1 1 1 1
1 1 1 1 1     1 2 2 2 1     1 2 3 2 1
1 1 1 1 1     1 2 3 2 1     1 3 5 3 1
1 1 1 1 1     1 2 2 2 1     1 2 3 2 1
1 1 1 1 1     1 1 1 1 1     1 1 1 1 1
```

The users can also design their own filter and enter it as a data file. This user-defined filter data file must have the same definition and format as those used by *FILTERSIM*, see Section 8.2.2 and Fig. 8.30. Notice that the number of filters (first line in the filter data file) must be exactly 1, and the filter weights can be any real values.

Given a K-categorical variable Z with current proportions p_k^c ($k = 1, \ldots, K$), a J-node filter $F = \{(\mathbf{h}_j, f_{\mathbf{h}_j}), j = 1, \ldots, J\}$, and target proportions p_k^t ($k = 1, \ldots, K$), the *TRANSCAT* algorithm proceeds as follows.

- At each node $\mathbf{u}$ of the grid to be processed, extract the local pattern $pat(\mathbf{u})$ within the moving window W.
- Calculate the local category pseudo proportions $p_k(\mathbf{u}), k = 1, \ldots, K$ of that pattern $pat(\mathbf{u})$ as

$$p_k(\mathbf{u}) = \frac{1}{N_W} \sum_{j \in \{N_W^k\}} \omega \cdot f_{\mathbf{h}_j} \qquad (9.3)$$

where:

$\{N_w^k\}$ is the set of nodes in window W which are informed as category k,
$N_W = \sum_{k=1}^{K} |N_W^k| \leq J$ is the total number of nodes informed in $pat(\mathbf{u})$,
$\omega = 1$ if node $(\mathbf{u} + \mathbf{h}_j)$ is not a hard data location; otherwise $\omega = f_0$, where
f_0 is an input control parameter giving more weight to hard data locations. The default value is $f_0 = 10$.

- Calculate the relative category pseudo proportions as

$$p_k^{\text{r}}(\mathbf{u}) = p_k(\mathbf{u}) \cdot \left(\frac{p_k^{\text{t}}}{p_k^{\text{c}}}\right)^{f_k}, \quad k = 1, \ldots, K \tag{9.4}$$

where $f_k \geq 1$ is an input factor associated with category k. The greater f_k, the more importance is given to the reproduction of target proportion p_k^{t}.
- Find the category k^F which maximizes p_k^{r}.
- Update the property at location $\mathbf{u}$ into $Z^{\text{new}}(\mathbf{u}) = k^F$.

Note that for the default filter $f_k = 1$, $(k = 1, \ldots, K)$, *TRANSCAT* acts as a de-noise program to clean a pixel-based image (Schnetzler, 1994).

Parameters description

The *TRANSCAT* algorithm is activated from *Utilities* → *transcat* in the Algorithm Panel. The *TRANSCAT* interface is given in Fig. 9.2, and the parameters

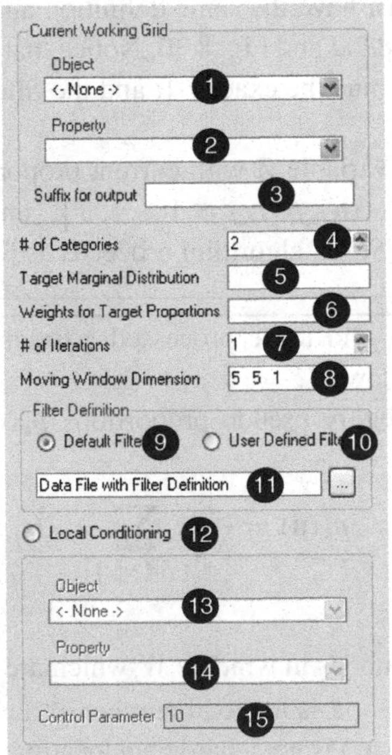

Figure 9.2 User interface for *TRANSCAT*

are described in the following. The text inside "[ ]" is the corresponding keyword in the *TRANSCAT* parameter file.

1. **Object** [Working_Grid.grid]   The name of the grid to be transformed.
2. **Property** [Working_Grid.property]   The name of the property to be transformed.
3. **Suffix for output** [Out_Suffix]   The name for the output property consists of the original name plus the suffix entered here.
4. **# of Categories** [Nb_Categories]   The number K of categories contained in the working property (item 2).
5. **Target Marginal Distribution** [Marginal_Pdf]   The target categorical proportions must be given in sequence from category 0 to Nb_Categories-1. The sum of all target proportions must be 1.
6. **Factors for Target Proportions** [Marginal_Pdf_Factor]   Relative factor f_k associated with each category in Eq. (9.4). The larger the factor, the greater control on its proportion adjustment. The total number of weights to be input is Nb_Categories.
7. **# of Iterations** [Nb_Iterations]   The number of iterations to perform *TRANSCAT* algorithm, the default value is 1.
8. **Moving Window Dimension** [Moving_Window]   The size of a 3D template to retrieve the local image patterns. The three input numbers must be positive odd integers, and separated by a space. This template is used to retrieve the required user-defined filter values.
9. **Default Filter** [Filter_Default]   The option to use the default filters provided by *TRANSCAT*.
10. **User Defined Filter** [Filter_User_Define]   The option to use the user's own filter definition. Note that 'Default' and 'User Defined' are mutually exclusive.
11. **Data File with Filter Definition** [User_Def_Filter_File]   Input a data file with the filter definitions (see Fig. 8.30), same as for *FILTERSIM*. This parameter is ignored if Filter_User_Define is set to 0.
12. **Local Conditioning** [Is_Cond]   Enables a transformation honoring local hard data.
13. **Object** [Cond_Data.grid]   The name of the grid containing the local hard data.
14. **Property** [Cond_Data.property]   The name of the local hard data property.
15. **Control Parameter** [Weight_Factor]   Relative weight f_0 defined in formula Eq. (9.3), and used for the hard conditioning data; must be larger than the weights f_k given in item **6** to privilege the impact of hard data.

9.3 POSTKRIGING: post-processing of kriging estimates

Some applications require more displays than just the kriging estimate or kriging variance. One may be interested in knowing the probability to be above or below a threshold value for example. Algorithm *POSTKRIGING* builds at each location a probability distribution for the unknown and retrieves information from it. In the case of indicator kriging, a non-parametric ccdf is built from the vector of estimated probabilities of being below the input threshold values, with the tail distributions defined by the user using the SGeMS parametrization presented in Section 6.8. From that ccdf, the conditional mean and variance as well as quantiles, probability of being above or below a threshold and the interquartile range can be retrieved. The same information can be retrieved from the results of kriging or cokriging a continuous variable, by identifying the kriging estimate and the kriging variance with the mean and the variance of a Gaussian distribution. If there is no reason to believe the variable follows a Gaussian distribution, estimation from indicator kriging should be preferred to retrieve probability-type information.

Parameters description

The *POSTKRIGING* algorithm is activated from *Utilities* → *postKriging* in the Algorithm Panel. The main *POSTKRIGING* interface contains two pages: "Distribution" and "Statistics" (see Fig. 9.3). The text inside "[ ]" is the corresponding keyword in the *POSTKRIGING* parameter file.

1. **Grid Name** [Grid_Name]   Name of the grid.
2. **Distribution type**   Either choose a non-parametric distribution [is_non_param_cdf] or Gaussian distribution [is_Gaussian].
3. **Properties with probabilities** [props]   Properties containing the probabilities to be below given thresholds. The results from *INDICATOR KRIGING* yields directly such properties.
4. **Thresholds** [marginals]   Thresholds associated with each of the properties.
5. **Lower tail extrapolation** [lowerTailCdf]   Parametrization of the lower tail.
6. **Upper tail extrapolation** [upperTailCdf]   Parametrization of the upper tail.
7. **Gaussian Mean** [gaussian_mean_prop]   Property containing the local mean of a Gaussian distribution, e.g. obtained from kriging.
8. **Gaussian Variance** [gaussian_var_prop]   Property containing the local variance of a Gaussian distribution, e.g. obtained from kriging.
9. **Mean**   If [mean] is selected then the conditional mean is computed locally from the local distribution and output in the property [mean_prop].

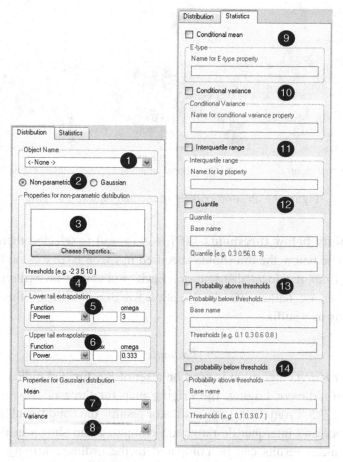

Figure 9.3 User interface for *POSTKRIGING*

10. **Variance** If [cond_var] is selected then the conditional variance is computed from the local distribution and output in the property [cond_var_prop].

11. **Interquartile range** If [iqr] is selected then the interquartile range ($q_{75} - q_{25}$) is computed and output in the property [iqr_prop].

12. **Quantile** If [quantile] is selected then the quantile for the specified probability values in [quantile_vals] are computed from the local distribution. Each quantile is written in a property with base name given by [quantile_prop] with the quantile value appearing as suffix.

13. **Probability above thresholds** If [prob_above] is selected then the probability to be above the thresholds specified in [prob_above_vals] are computed from the local distribution. Each probability is written in a property with base name given by [prob_above_prop] with the threshold value appearing as a suffix.

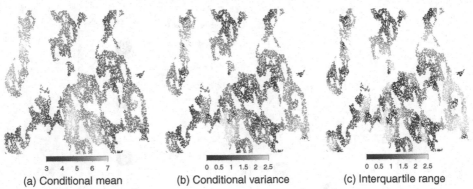

| 3 4 5 6 7 | 0 0.5 1 1.5 2 2.5 | 0 0.5 1 1.5 2 2.5 |
| (a) Conditional mean | (b) Conditional variance | (c) Interquartile range |

Figure 9.4 Post-processing of median indicator kriging

14. Probability below thresholds If [prob_below] is selected then the probability to be below the thresholds specified in [prob_below_vals] is computed locally from the local distribution. Each probability is written in a property with base name given by [prob_below_prop] with the threshold value appearing as a suffix.

Example

The *POSTKRIGING* algorithm is used on a set of 8 estimated indicators obtained from median indicator kriging; the kriging map for three of these indicators is shown in Fig. 7.5. From the kriging indicator results, the conditional mean, conditional variance and interquartile range are computed and shown in Fig. 9.4. The conditional mean results can be compared to the ordinary kriging map shown in Fig. 7.3. Particularly interesting is the difference between the estimation variances resulting from simple kriging and from the indicator kriging approach. The inter-quartile range provides a different measure of uncertainty from the variance.

9.4 *POSTSIM*: post-processing of realizations

The *POSTSIM* algorithm extracts local statistics from a set of simulated realizations. For each node in a grid, a probability distribution is built from the realizations available, then user-specified statistics such as mean, variance, interquartile range, probability to be above or below a threshold, and the mean above or below a threshold, can be computed. Each realization provides a structurally accurate map, i.e. a map that reproduces the spatial patterns implicit to a variogram model or extracted from a training image. The point-wise averaging of multiple equiprobable realizations provides a single so-called E-type map with local accuracy similar to that of the corresponding kriging map.

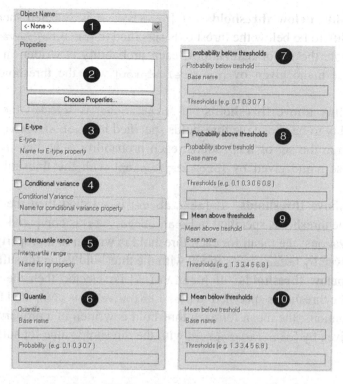

Figure 9.5 User interface for *POSTSIM*

Parameters description

The *POSTSIM* algorithm is activated from *Utilities* → *postSim* in the Algorithm Panel. The text inside "[ ]" is the corresponding keyword in the *POSTSIM* parameter file. The user interface is shown in Fig. 9.5.

1. **Object Name** [Grid_Name]   Name of the working Cartesian grid or point-set.
2. **Properties** [props]   Realizations to be processed.
3. **E-type**   If [mean] is selected then the local conditional mean is computed from the set of realizations and output in the property [mean_prop].
4. **Variance**   If [cond_var] is selected then the local conditional variance is computed from the set of realizations and output in the property [cond_var_prop].
5. **Interquartile range**   If [iqr] is selected then the local interquartile range is computed from the set of realizations and output in the property [iqr_prop].
6. **Quantile**   If [quantile] is selected then the local quantile for the values specified in [quantile_vals] is computed from the set of realizations; each quantile is written in a property with base name given by [quantile_prop] with the quantile value for suffix.

7. **Probability below thresholds**   If [prob_below] is selected then the local probability to be below the thresholds specified in [prob_below_vals] is computed from the set of realizations; each probability is written in a property with base name given by [prob_below_prop] with the threshold value for suffix.

8. **Probability above thresholds**   If [prob_above] is selected then the local probability to be above the thresholds specified in [prob_above_vals] is computed from the set of realizations; each probability is written in a property with base name given by [prob_above_prop] with the threshold value for suffix.

9. **Mean above thresholds**   If [mean_above] is selected then the local mean above the thresholds specified in [mean_above_vals] is computed from the set of realizations; the mean for each threshold is written in a property with base name given by [mean_above_prop] with the threshold value for suffix.

10. **Mean below thresholds**   If [mean_below] is selected then the local mean below the thresholds specified in [mean_below_vals] is computed from the set of realizations; the mean for each threshold is written in a property with base name given by [mean_below_prop] with the threshold value for suffix.

Example

The *POSTSIM* algorithm is used to get the conditional mean (E-type), variance and interquartile range of a set of 50 realizations as generated by the direct sequential simulation algorithm. The *POSTSIM* results, shown in Fig. 9.6, are to be compared with the kriging results (Fig. 7.3) and the post-processing of indicator kriging (Fig. 9.4a).

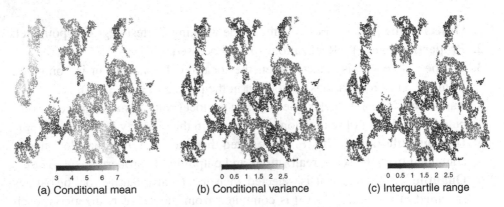

(a) Conditional mean   (b) Conditional variance   (c) Interquartile range

Figure 9.6 Post-processing of 50 *DSSIM* realizations

9.5 *NU-TAU MODEL*: combining probability fields

Combining sources of information can be done using the nu/tau model described in Section 3.10. The nu/tau model allows to merge N elementary probabilities, each conditional to a single datum event, into a single posterior probability conditional to all data events considered jointly. The nu/tau parameters account for information redundancy between these data events.

The SGeMS implementation of the *NU-TAU MODEL* allows to enter the redundancy either with the tau (τ) or the nu (ν) parameters. Recall that the conversion between ν and τ is data dependent.

The N ν parameters can be lumped into a single product, the ν_0 parameter. By default ν_0 is set to one which corresponds to no data redundancy. SGeMS offers three ways to enter the redundancy correction: (a) enter a single ν_0 value, (b) enter one ν or τ value for each source of information, and (c) enter a property that contains a ν_0 value at each location or similarly a vector of properties with the τ_i values. Options (a) and (b) amount to assuming the same redundancy model at all locations. Option (c) allows to define a location-specific redundancy model. Recall that the ν_i and the ν_0 values must be positive.

Wherever a location has contradictory information, e.g. one source of information gives a probability of one while a second source gives a probability of zero, that location is left uninformed and a warning message is issued.

Parameters description

The *NU-TAU MODEL* algorithm is activated from *Utilities* $\rightarrow$ *nutauModel* in the Algorithm Panel. The text inside "[ ]" is the corresponding keyword in the *NU-TAU MODEL* parameter file. The user interface is shown in Fig. 9.7.

1. **Object Name** [Grid]   Name of the grid.
2. **Properties** [props]   Properties containing the single event conditional probability. More than one property must be selected.
3. **Properties** [nu_prop]   Name for the output property: the combined conditional probability.
4. **Redundancy Model**   Model selection: select [is_tau] for the tau model or [is_nu] for the nu model.
5. **Redundancy Value**   Indicate if the ν or τ are constant ([redun_constant] is selected) or location specific ([redun_specific] is selected). In the former case, the user can either enter the single value ν_0 or one ν value for each probability map. In the latter case, the ν_0 are stored on a property. Only needed if [is_nu] is selected.

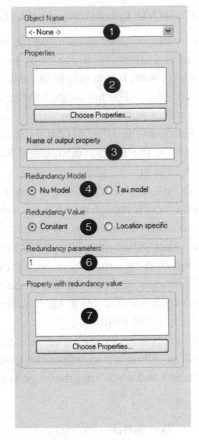

Figure 9.7 User interface for *NU-TAU MODEL*

6. **Redundancy parameters** [redun_input]   Required if [redun_constant] is selected. The τ or the ν are entered for each property selected in [props]. For the nu model, if a single value is entered it will be considered as ν_0.

7. **Redundancy property** [redun_prop]   Required if [redun_specific] is selected. Select a property containing the ν_0 value for each location of the grid when [is_nu] is selected, or a vector of properties with the τ_i parameters when [is_tau] is selected. Uninformed nodes are skipped.

9.6 *BCOVAR*: block covariance calculation

Given a variogram/covariance model and a gridded field, it is useful to visualize the spatial correlation between any point or block and other point in the field; this is provided by the point-to-point or block-to-point covariance map. It is also helpful to calculate and display the covariance value between any two given points or blocks. *BCOVAR* is the utility program to achieve that. It allows calculation of the types of covariances or covariance maps required by the kriging equations using block data, see Section 3.6.3 and Eq. (7.10).

Block-to-point covariance map

In *BCOVAR*, the covariance between any given block average and any point node within a gridded field is computed. Either the traditional integration method or the integration-hybrid method can be used. A property map or cube associated with the grid is generated that holds the calculated covariance values; such a map is easy to visualize. Also, those covariance values can be saved into a file through SGeMS menu: *Objects | Save Object.*

Point-to-point covariance map

The covariance map between any node and the gridded field is calculated. The covariance values are stored on a property map for visualization. Again, these can be saved into a file through SGeMS menu: *Objects | Save Object.*

Block-to-block, block-to-point and point-to-point covariance value

BCOVAR allows to compute one single covariance value related to block-to-block, block-to-point or point-to-point. This value is shown on a pop-up window. Such a covariance value provides a quantitative measure of spatial correlation between the two specified blocks or points.

Parameters description

The *BCOVAR* algorithm is activated from *Utilities* → *bcovar* in the Algorithm Panel. The main *BCOVAR* interface contains two pages: "General" and "Variogram" (see Fig. 9.8). The text inside "[ ]" is the corresponding keyword in the *BCOVAR* parameter file.

1. **Grid Name** [Grid_Name]   Name of the simulation grid.
2. **Block Covariance Computation Approach** [Block_Cov_Approach]   Select the method of computing block covariance: **FFT with Covariance-Table** or **Integration with Covariance-Table**.
3. **Covariance Calculation Type** [Cov_Cal_Type]   Select the type of covariance for computation: **Block-to-Point Cov. Map, Point-to-Point Cov. Map, Block-to-Block Cov. Value, Block-to-Point Cov. Value** and **Point-to-Point Cov. Value.**
4. **Block-to-Point Cov. Map Parameters**   Only activated if **Covariance Calculation Type** [Cov_Cal_Type] is set to **Block-to-Point Cov. Map**. The block

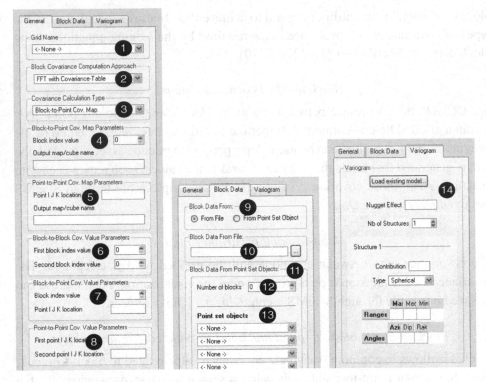

Figure 9.8 User interface for *BCOVAR*

used is entered by its block ID through the parameter **Block index value**
[Block_Index_for_Map]. All blocks are internally assigned a block ID from 0
to $N - 1$ based on their locations in the block data file, where N is the number
of blocks. The output block covariance map/cube property name is specified
through **Output map/cube name** [Block_Cov_Output_Prop_Name].

5. **Point-to-Point Cov. Map Parameters** Only activated if **Covariance Calculation Type** [Cov_Cal_Type] is set to **Point-to-Point Cov. Map**. The
point used is entered by its $i j k$ index through the parameter **Point I
J K location** [Point_IJK_for_Point_Cov_Map]. Note that all indexes start
from 0. The input $i j k$ are separated by blanks. The output point covariance map/cube property name is specified through **Output map/cube name**
[Point_Cov_Output_Prop_Name].

6. **Block-to-Block Cov. Value Parameters** Only activated if **Covariance Calculation Type** [Cov_Cal_Type] is set to **Block-to-Block Cov. Value**. The two
blocks used are entered by its block index through the parameters **First
block index value** [First_Block_Index] and **Second block index value**
[Second_Block_Index]. Note that all blocks are internally assigned a block

index from 0 to $N - 1$ based on their input sequence in the block data file, where N is the number of blocks. The calculated covariance value is shown on a pop-up window at the end of the calculation.

7. **Block-to-Point Cov. Value Parameters** Only activated if **Covariance Calculation Type** [Cov_Cal_Type] is set to **Block-to-Point Cov. Value**. The block used is entered by its block index through the parameter **Block index value** [Block_Index_for_BP_Cov] and the point used is entered by its $i j k$ index through **Point I J K location** [Point_IJK_for_BP_Cov]. Note that all blocks are internally assigned a block index from 0 to $N - 1$ based on their input sequence in the block data file, where N is the number of blocks. All point indexes start from 0. The input $i j k$ are separated by blanks. The calculated covariance value is shown on a pop-up window at the end of the calculation.

8. **Point-to-Point Cov. Value Parameters** Only activated if **Covariance Calculation Type** [Cov_Cal_Type] is set to **Point-to-Point Cov. Value**. The two points used are entered by their indexes through the parameters **First point I J K location** [First_Point_IJK] and **Second point I J K location** [Second_Point_IJK]. Note that all indexes start from 0. The input $i j k$ are separated by blanks. The calculated covariance value is shown on a pop-up window at the end of the calculation.

9. **Block Data From** Select where the block data are to be found. There are two options: **From File** [Block_From_File] and **From Point Set Object** [Block_From_Pset].

10. **Block Data From File** [Block_Data_File] Only activated if **From File** [Block_From_File] is selected in **Block Data From**. The directory address of the block data file should be specified. The block data file format is shown in Fig. 7.9. If no block data file is entered, the estimation is performed using only point data.

11. **Block Data From Point Set Objects** Only activated if **From Point Set Object** [Block_From_Pset] is selected in **Block Data From**.

12. **Number of blocks** [Number_of_Blocks] Number of blocks entered from point-set objects.

13. **Point set objects** [Block_Grid_i] Enter the point-set block grids. This allows users to conveniently use the pre-loaded point-set objects. No property is required to be associated with the input point-set grids. The maximum number of grids entered in this way is 50. They have to be loaded from a file if there are more than 50 blocks. The number of input point-set grids should be equal to or larger than the **Number of blocks** [Number_of_Blocks].

14. **Variogram** [Variogram_Cov] Parametrization of the variogram model, see Section 6.5.

Examples

BCOVAR is run on a Cartesian grid of 40×40 with grid dimension of 0.025×0.025. The variogram model (7.11) is used.

The block-to-point covariance map of the block (whose geometry is given in Fig. 9.9a) is generated, see Fig. 9.9b. The covariance value is higher close to the block location and decreases away from the block.

The covariance map of a point (15, 25, 0) (see location in Fig. 9.9c) is generated, see Fig. 9.9d. At the point location, the covariance value is 1 and decreases away from it. The anisotropy character of the input variogram model is revealed. The covariance value between the block and the point (see Fig. 9.9e) is calculated and the result is shown on the pop-up window (Fig. 9.9f).

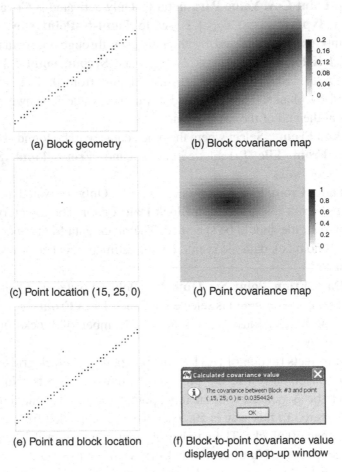

(a) Block geometry   (b) Block covariance map

(c) Point location (15, 25, 0)   (d) Point covariance map

(e) Point and block location   (f) Block-to-point covariance value
                                          displayed on a pop-up window

Figure 9.9 Covariance map or value obtained from *BCOVAR*

9.7 *IMAGE PROCESSING*

The *IMAGE PROCESSING* algorithm allows one to scale and rotate properties on a Cartesian grid. It is especially useful for mp simulations either with *SNESIM* or *FILTERSIM*. For instance, geological structures from the available training image may not have the desired thickness or orientation. *IMAGE PROCESSING* allows to process that imperfect training image to provide more relevant structural information to the mps simulation algorithms. With the rotation and scaling operations, the *IMAGE PROCESSING* program can modify both the geobodies' directions and size but not their topology.

Given the central location $\mathbf{u}_0^s$ of the original training image (source), the central location $\mathbf{u}_0^t$ of the new training image (target), the three rotation angles α, β, θ, and the three affinity factors f_x, f_y, f_z, each node $\mathbf{u}^t$ in the new training image (Ti^{new}) has the same value at node $\mathbf{u}^s$ in the original training image (Ti^{old}), here $\mathbf{u} = (x, y, z)'$. The relationship between $\mathbf{u}^t$ and $\mathbf{u}^s$ is given by

$$\mathbf{u}^t = \mathbf{T} \cdot \Lambda \cdot (\mathbf{u}^s - \mathbf{u}_0^s) + \mathbf{u}_0^t , \qquad (9.5)$$

where the rotation matrix $\mathbf{T}$ is defined as:

$$\mathbf{T} = \begin{bmatrix} \cos\theta & 0 & -\sin\theta \\ 0 & 1 & 0 \\ \sin\theta & 0 & \cos\theta \end{bmatrix} \begin{bmatrix} 1 & 0 & 0 \\ 0 & \cos\beta & \sin\beta \\ 0 & -\sin\beta & \cos\beta \end{bmatrix} \begin{bmatrix} \cos\alpha & \sin\alpha & 0 \\ -\sin\alpha & \cos\alpha & 0 \\ 0 & 0 & 1 \end{bmatrix}$$

and the scaling matrix Λ is defined as

$$\Lambda = \begin{bmatrix} f_x & 0 & 0 \\ 0 & f_y & 0 \\ 0 & 0 & f_z \end{bmatrix}.$$

The three rotation angles α, β, θ are the azimuth, dip and plunge, see Section 2.5.

Note that the rotation sequence is important: rotation is performed first around the Z-axis, then around the X-axis, and finally around the Y-axis. The larger the input affinity factor, the larger the geobody in the output training image.

Parameters description

The *IMAGE PROCESSING* algorithm is activated from *Utilities* → *ImageProcess* in the Algorithm Panel. The *IMAGE PROCESSING* interface is given in Fig. 9.10, and the parameters are described in the following. The text inside "[ ]" is the corresponding keyword in the *IMAGE PROCESSING* parameter file.

1. **Source Grid** [Source_Grid]   The object holding the training image property to be modified. It must be a Cartesian grid.

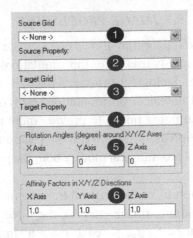

Figure 9.10 User interface for *IMAGE PROCESSING*

2. **Source Property** [Source_Property]   The original training image property to be modified.

3. **Target Grid** [Target_Grid]   The object that holds the modified training image property must be a Cartesian grid.

4. **Target Property** [Target_Property]   The new training image property name after modification.

5. **Rotation Angles (degree) around X/Y/Z Axes**   The input rotation angles around each X/Y/Z axis; these angles must be given in degree units, see Section 2.5 for a description of the rotation angles. The three keywords are: [Angle0], [Angle1] and [Angle2].

6. **Affinity Factors in X/Y/Z Directions**   The input affinity factors along each X/Y/Z direction, these scaling factors must be positive. The larger the input affinity factor, the larger the structures in the output property. The three keywords are: [Factor0], [Factor1] and [Factor2].

9.8 *MOVING WINDOW*: moving window statistics

The *MOVING WINDOW* algorithm allows one to process both Cartesian grid and point-set grid with a moving window. At every location along a grid, statistics related to values within the window are computed and assigned to the window central point. A moving weighted linear average, or filter, is used in the *FILTER-SIM* algorithm where patterns are classified according to scores defined from such filters. Moving windows statistics are useful during exploratory data analysis to detect trends, or non-stationary regions within a grid.

The filters available in *MOVING WINDOW* are as follows.

Moving average Compute the linear average of the neighborhood.

Gaussian low pass Weighted linear average with weights given by a Gaussian function with maximum value at the center pixel and decreasing away from the center. The weights are given by

$$\lambda(\mathbf{h}) = \frac{1}{\sigma} \exp\left(\frac{-\parallel \mathbf{h} \parallel^2}{\sigma}\right)$$

where σ is a user-defined parameter controlling the rate of decrease of the weights with distance $\parallel \mathbf{h} \parallel$.

Moving variance Compute the variance of the neighborhood.

Default filtersim filters These filters are the same as used in the *FILTERSIM* algorithm described in Section 8.2.2.

Sobel filters Edge detection filters; the intensity and direction of edges can also be retrieved. This is a 2D filter; when applied to 3D grids, it will process the full volume one layer or section at a time. The user can specify the plan in which the Sobel filter is to be applied (XY, XZ or YZ)

$$G_x = \begin{bmatrix} -1 & 0 & 1 \\ -2 & 0 & 2 \\ -1 & 0 & 1 \end{bmatrix} \qquad G_y = \begin{bmatrix} 1 & 2 & 1 \\ 0 & 0 & 0 \\ -1 & -2 & -1 \end{bmatrix}.$$

From the Sobel filter's output, the intensity of an edge G is given by $G = \sqrt{G_x^2 + G_y^2}$ and its orientation by $\theta = \tan^{-1}(G_y/G_x)$.

User-defined filters Filters with weights and shape fully defined by the user. They have the same definition and format as with *FILTERSIM*, see Section 8.2.2 and Fig. 8.30.

With the exception of the Sobel filter, all filters are 3D with user-specified window dimension. Grid edge pixels are not processed and are left uninformed.

Tip 8 Canny segmentation

The Canny segmentation method (Canny, 1986) is an edge detection technique widely used in image processing that can be easily implemented in SGeMS. It consists of first applying a Gaussian low pass filter followed by the Sobel filter.

Tip 9 Categorical variables

The filters defined above are designed for a continuous attribute but can also be applied on binary variables. In case of a categorical attribute, we suggest transforming these categories into a set of indicator variables on which these filters can be applied.

Parameters description

The *MOVING WINDOW* algorithm is activated from *Utilities* → *MovingWindow* in the Algorithm Panel. The main *MOVING WINDOW* interface is given in Fig. 9.11, and the parameters are described in the following. The text inside "[ ]" is the corresponding keyword in the *MOVING WINDOW* parameter file.

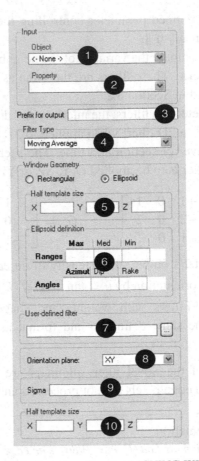

Figure 9.11 User interface for *MOVING WINDOW*

1. **Input—Object** [Input_data.grid]   Grid containing the property to be processed.
2. **Input—Property** [Input_data.property]   Property to be processed.
3. **Prefix** [prefix_out]   Prefix to be added to the input property name to identify the output property.

4. **Filter Type** [filter_type]   Selection of the filter; the choices are: moving average, moving variance, default *FILTERSIM* filter, Gaussian low pass, Sobel edge detection filter or the user-defined option.

5. **Half template size**   This entry parametrizes a rectangular moving window. If [grid] is a Cartesian grid, [size_x] [size_y] and [size_z] correspond to the number of template pixels on each side of the center pixel, (e.g. the total size of pixel in x would be $2 \times$ [size_x] $+ 1$). A Cartesian grid requires integer values. If [grid] is a point-set then the entries correspond to the radius of an ellipsoid.

6. **Ellipsoid definition**   This entry parametrizes an ellipsoid moving window.

7. **User-defined filter** [filter_filename]   Load the user-defined filters with the same format as in *FILTERSIM*, see Section 8.30.

8. **Plane for the Sobel filter**   The Sobel filter is only 2D, and can be applied in the XY, XZ or YZ planes by selecting [plan_xy], [plan_xz] or [plan_yz] respectively.

9. **Sigma** [sigma]   The variance of the Gaussian filter, only available when the "Gaussian low pass" option is selected, see item 4.

10. **Half template size**   The size of the *FILTERSIM* filters, see Section 8.30.

9.9 *TIGENERATOR*: object-based image generator

The multiple-point simulation algorithms *SNESIM* and *FILTERSIM* require, as input structural model, a training image (Ti) that is a numerical description of the perceived geological heterogeneity. This training image is a conceptual representation of the type of structures deemed present in the field/reservoir. Unconditional realizations of object-based algorithms such as *fluvsim* (Deutsch and Tran, 2002) and *ellipsim* (Deutsch and Journel, 1998) have been used for generating Tis with channel and elliptical geometries respectively. However, owing to the lack of a common interface, modeling interactions between the different objects is limited to simple overlap of the generated objects.

TIGENERATOR provides a single program to generate different parametric shapes using non-iterative, unconditional Boolean simulation. In addition, any non-parametric shape can be simulated by specifying a rasterized template of that shape. *TIGENERATOR* is fast because it is unconditional. It is flexible because the pattern distribution of a variety of depositional environments can be modeled using the available shapes and constraints. The various object geometry parameters and orientation can be kept constant or made to follow pre-defined probability distributions. Interaction between objects can be modeled by specifying erosion and overlap rules.

The structure of the *TIGENERATOR* program is outlined in Algorithm 9.2.

Algorithm 9.2   Training image generator

1: Initialize a list of geobodies and their input proportions
2: **for** Each geobody i in the list **do**
3:     Define a random path visiting each node of the grid
4:     **for** Each node **u** along the path **do**
5:         Draw geobody i
6:         Update the proportion of geobody i
7:         **if** Simulated proportion $\geq$ target proportion of geobody i **then**
8:             Break out of loop
9:         **end if**
10:    **end for**
11: **end for**
12: Repeat for another realization

The current implementation provides four parametric shapes: sinusoid, ellipsoid, half-ellipsoid, and cuboid.

Sinusoid In the horizontal plane a sinusoid is parametrized by an amplitude and a wavelength (Fig. 9.12a). Its vertical cross-section is a lower half-ellipse, which is described by a width and a thickness (Fig. 9.12b). The sinusoid object can be rotated only in the horizontal plane.

Ellipsoid The ellipsoid object is parametrized by a maximum, medium and minimum axis exactly like the search ellipsoid of Section 2.5. It can be rotated only in the horizontal plane.

Half-ellipsoid The vertical upper or lower section of a full ellipsoid. It can be rotated only in the horizontal plane.

Cuboid A rectangular parallelepiped characterized by its X, Y and Z dimensions. It can be rotated in both the horizontal and vertical planes.

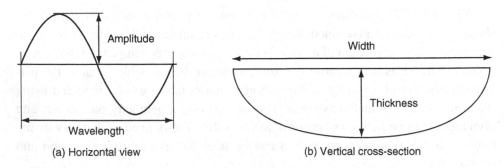

(a) Horizontal view   (b) Vertical cross-section

Figure 9.12 Horizontal and vertical views of a sinusoid

The object geometry and orientation parameters can be made to follow independently three types of distributions: Dirac, uniform or triangular.

Constant value or Dirac distribution Completely defined by the single parameter [Mean]; it is used to input constant parameters.

Uniform distribution A uniform distribution is parametrized with [Min] and [Max], the lower and the upper bounds of the distribution.

Triangular distribution A triangular distribution is parametrized with [Min], [Mode] and [Max]: the lower bound, the mode and the upper bound of the distribution.

9.9.1 Object interaction

A training image must not only depict each object, but also how these objects spatially relate to each other. Geological events cause objects to overlap and erode one another. In *TIGENERATOR* these spatial relationships between objects are modeled by user-specified erosion and overlap rules.

Erosion rules: these must be specified between a geobody and all geobodies that are simulated prior to it. A geobody can either erode (code 1) or be eroded by (code 0) the previously simulated geobodies. All geobodies erode the background by default.

Overlap rules: these help constrain the fraction of volumetric overlap between two geobodies. The overlap is controlled by two parameters, minimum overlap and maximum overlap, which are bounded by [0, 1]. For each geobody, the user must specify an overlap rule with all previous geobodies. If the volumetric overlap between two geobodies is to be the same, then minimum and maximum overlap should be set equal.

To illustrate some end member overlap cases, consider the example of a three facies training image – background, ellipsoid and cuboid – simulated in that order. Prior to any geobody simulation all grid nodes are set to background facies. The ellipsoid object is simulated first. No erosion or overlap rules need to be specified for the ellipsoid object because it is simulated first. The cuboid objects, which are simulated next (second), erode the ellipsoids. Consider the following four cases of overlap between the cuboid and ellipsoid objects: if, for the cuboid object, min overlap and max overlap are both set to 0, i.e. 0% volumetric overlap, then the two geobodies will not be in contact (Fig. 9.13a). On the other hand, if min and max overlap are both set to 1, then cuboids will be completely contained within the ellipsoids (Fig. 9.13b). Setting min overlap to 0.01 will ensure that cuboids are attached with at least a 1% volumetric overlap to ellipsoids (Fig. 9.13c). If no

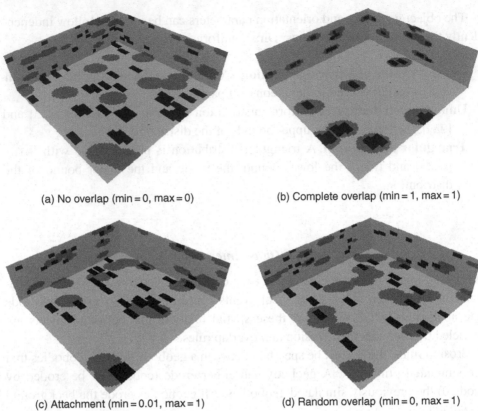

(a) No overlap (min = 0, max = 0)   (b) Complete overlap (min = 1, max = 1)

(c) Attachment (min = 0.01, max = 1)   (d) Random overlap (min = 0, max = 1)

Figure 9.13 Examples of interaction between ellipsoid and cuboid objects using different values of minimum and maximum overlap values. The values specified are for the cuboid object

specific overlap relationship exists between the two geobodies then set min overlap to 0 and max overlap to 1, which results in a random overlap (Fig. 9.13d).

Warning: it is the user's responsibility to ensure consistency between the different overlap rules. If one or more geobodies cannot be simulated owing to inconsistent overlap criteria, the simulation can be stopped by clicking Abort on the progress dialog.

Parameters description

The *TIGENERATOR* algorithm is activated from *Utilities* → *TiGenerator* in the Algorithm Panel. The main *TIGENERATOR* interface is given in Fig. 9.14, and the parameters are described below. The text inside "[ ]" is the corresponding keyword in the *TIGENERATOR* parameter file.

1. **Simulation Grid Name** [ti_grid]   The name of the simulation grid.
2. **Property Name Prefix** [ti_prop_name]   Prefix for the simulation output. The suffix _real# is added for each realization.
3. **Nb of realizations** [nb_realizations]   Number of simulations to generate.
4. **Seed** [seed_rand]   Seed for the random number generator (should be a large odd integer).
5. **Nb of geobodies** [nb_geobodies]   Number of different geobody types to be simulated. Minimum value of this parameter is set to 1 (at least one geobody must be simulated). Total number of facies in the Ti is equal to the number of geobodies plus the background facies (index 0).
6. **Geobody information**   Information about a geobody is collected using a geobody selector [geobodySelector]. For each geobody the following input is required: index [gbIndex], geobody type [gbType], geobody parameters, proportion [gbProp], and interaction rules with all previously simulated geobodies. Note that the geobodies are simulated in the order in which they are specified; geobody indexing should be sequential (1, 2, 3, etc). Select one geobody type from the drop-down menu and click on the adjacent push button to invoke the corresponding geobody dialog. The geobody parameters can either be set as constant or specified as a uniform or triangular distribution. All dimensions should be specified in number of cells. Rotation for all geobodies, except the cuboid, is limited to the horizontal plane and the 3D search ellipsoid convention is used, see Section 2.5. The cuboid object can be rotated in both the horizontal (azimuth) and vertical (dip) planes. The types of geobodies available in *TIGENERATOR* and a description of their parameters are as follows.

   Sinusoid   The dimensions of the sinusoid object are determined by three parameters: length [sinLen], width [sinWid] and thickness [sinThk]. The horizontal sinuosity is specified by an amplitude [sinAmp] and a wavelength [sinWvl]. Horizontal orientation of the sinusoid [sinRot] is specified as clockwise rotation in degrees. No rotation implies that the sinusoid axis is North–South.

   Ellipsoid   The geometry of the ellipsoid object is determined by three radii: Max radius [ellipMaxr], Med radius [ellipMedr], and Min radius [ellipMinr]. Ellipsoid orientation [ellipRot] is specified as clockwise rotation in degrees from North (y-axis).

   Half-ellipsoid   The half-ellipsoid object can be either a lower half ellipsoid [lhellip] or an upper half ellipsoid [uhellip]. Its geometry is determined by three radii: Max radius [hellipMaxr], Med radius [hellipMedr], and Min radius [hellipMinr]. Half-ellipsoid orientation [hellipRot] is specified as clockwise rotation in degrees from North (y-axis).

Cuboid The cuboid object has three parameters: length [cubLen], width [cubWid] and height [cubHgt]. Cuboid orientation is specified through two angles: [cubRotStrike] is the clockwise rotation in the horizontal plane, and [cubRotDip] is dip rotation in degrees from the horizontal plane. When [cubRotStrike] is set to 0, the cuboid length runs North–South.

User-defined User-defined shape refers to any rasterized shape provided by the user through a text file [filename]. The file should contain three columns and as many lines as number of points in the raster. Each line contains, separated by spaces, the i, j, and k coordinates of a point in the raster relative to a center location (0, 0, 0). The shape need not be symmetric. Orientation [udefRot] is specified as clockwise rotation in degrees in the horizontal plane. The user-defined shape can be scaled by providing affinity factors in x, y, and z directions [udefScal]; these factors should be greater than zero.

For any given geobody, the interaction with previous geobodies must be specified by invoking the Interaction dialog shown in Fig. 9.14. Three types of interaction rules are required.

7. **Erosion rules with previous geobodies** The erosion rules are specified as a space-separated string of 0s and 1s. If the current geobody erodes a previous geobody, then the erosion value is 1, 0 otherwise. For the first geobody this field should be left blank as there are no previously simulated geobodies.

8. **Overlap with previous geobodies** The overlap rules control the fraction of volumetric overlap between two geobodies. It is specified as a pair of minimum [min] and maximum [max] overlap values, which are between [0, 1]. For the first geobody these fields should be left blank as there are no previously simulated geobodies. See Section 9.9.1 for details.

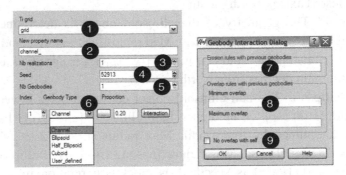

Figure 9.14 User interface for the *TIGENERATOR*

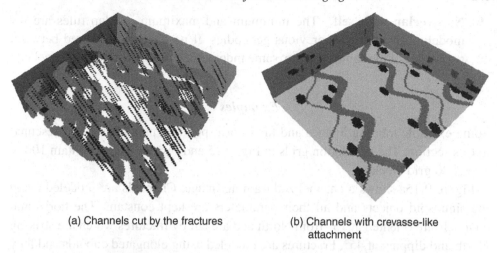

(a) Channels cut by the fractures

(b) Channels with crevasse-like attachment

Figure 9.15 Examples of parametric shapes generated using the *TIGENERATOR*

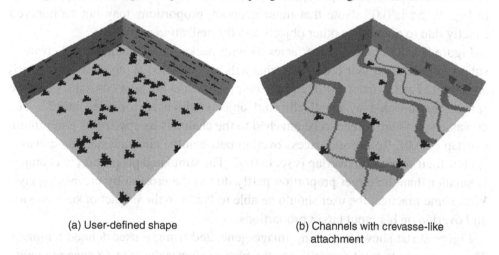

(a) User-defined shape

(b) Channels with crevasse-like attachment

Figure 9.16 Examples of training images containing both parametric and user-defined shapes

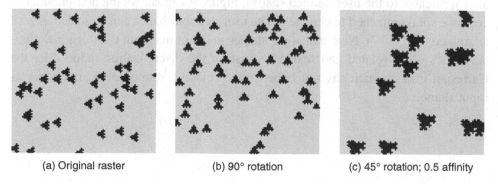

(a) Original raster

(b) 90° rotation

(c) 45° rotation; 0.5 affinity

Figure 9.17 Rotation and affinity of user-defined shapes

9. No overlap with self   The minimum and maximum overlap rules are for modeling overlap with previous geobodies. If no overlap is desired between geobodies of the same type (i.e. same index), then this box should be checked.

Examples

Some example training images and their corresponding parameters are presented in this section. The simulation grids in Fig. 9.15 and Fig. 9.16 each contain 100 × 100 × 30 grid blocks.

Figure 9.15a shows a channelized training image. Channels are modeled using the sinuosoid objects and all their parameters are kept constant. The horizontal channels are oriented at 15° from North and are cut by fractures which are striking North and dipping at 45°. Fractures are modeled using elongated cuboids and they erode the channels. The input channel proportion is 0.10; the simulated proportion in Fig. 9.15a is 0.09. Note that input geobody proportions may not be honored exactly due to erosion by other objects and discretization round-ups.

Figure 9.15b depicts a three facies Ti with background, channels and crevasse splays simulated in that order. The channels are modeled using sinusoids, which are oriented at 15° from North and all their parameters are kept constant. Crevasse splays are modeled by small ellipsoid objects which are oriented at 15°. The crevasse splays are forced to be attached to the channels by specifying a minimum overlap of 0.01. To prevent excess overlap between the channels and the crevasse splays, their maximum overlap is set to 0.02. The simulated proportion of channels is smaller than the target proportion partly due to the erosion by crevasse splays. With some practice, the user should be able to factor in the impact of such erosion and overlap on his input target proportions.

Figure 9.16a shows a training image generated using a user-defined template. This feature adds great versatility to the type of shapes that can be generated with the *TIGENERATOR*. All object interaction rules described in Section 9.9.1 are also applicable to the user-defined shapes. Figure 9.16b shows the non-parametric crevasse splays attached to channels. The user-defined shapes can be scaled as well as rotated (Fig. 9.17). Note that some shapes may not maintain their original characteristics when rotated and scaled because of the discretization induced by the Cartesian grid. In such cases it is better to provide a rotated or scaled raster as input shape.

10

Scripting, commands and plug-ins

It is often necessary to repeat similar tasks, e.g. to study the sensitivity of an algorithm to a given parameter. Such sensitivity analysis would be tedious if using the sole graphical interface, manually editing fields and clicking. SGeMS provides two ways to automate tasks: commands and Python scripts.

10.1 Commands

Most of the tasks performed in SGeMS using the graphical interface, such as creating a new Cartesian grid or performing a geostatistics algorithm, can also be executed using a command. A command is composed of a command name, e.g. NewCartesianGrid followed by a list of arguments delimited by "::". For example, the following command

```
NewCartesianGrid mygrid::100::100::10
```

creates a new Cartesian grid called mygrid, of dimensions $100 \times 100 \times 10$. To execute this command, type it in the command line editor, located on the Command Panel (see Fig. 10.1) and press Enter. If the Command Panel is not visible (it is hidden by default), display it by selecting *Commands Panel* from the *View* menu.

The SGeMS *Commands History* tab keeps a log of all commands performed either from the graphical interface or the command line. Executed commands appear in black. Messages, such as the time it took to execute the command, are displayed in blue, and warnings or errors in red. Commands can be copied from the *Commands History*, pasted in the command line, edited and executed.

The log of all SGeMS executed commands is also kept in a data file called sgems_history.log. Such a file is created in the directory from which SGeMS was started and contains all the commands executed during a SGeMS session, i.e. until SGeMS is exited. Note that the next time SGeMS is started, a new

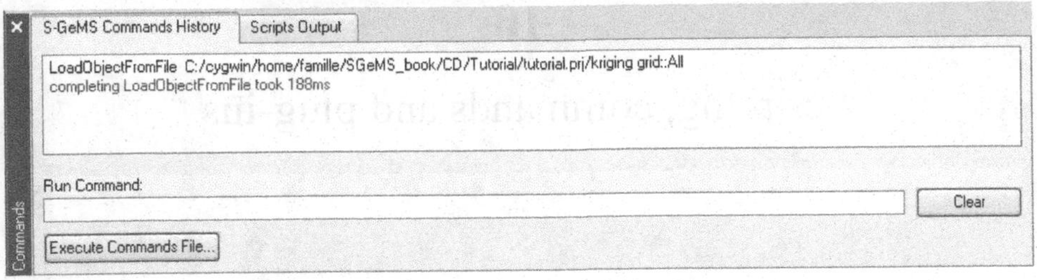

Figure 10.1 SGeMS Command Panel

`sgems_history.log` will be created, over-writing any existing log file with the same name.

Such log files allow one conveniently to record and replay a set of actions: to execute all the commands contained in a file, click the *Execute Commands File...* button and select the commands file. A commands file contains one command per line and lines that start with # are ignored.

10.1.1 Command lists

SGeMS understands many commands and it may not be easy to remember their names and expected parameters. The simplest way to recall a command name and its syntax is to execute the corresponding task using the graphical interface and copy the command from the History tab or the `sgems_history.log` file.

Another option is to use the Help command (just type `Help` in the command line editor). Help will list all available commands and briefly recall their syntax. The following is a list of all SGeMS commands with their input parameters. Parameters between square brackets are optional.

- `Help` List all the commands available in SGeMS.
- `ClearPropertyValueIf Grid::Prop::Min::Max` Set to not-informed all values of property `Prop` in grid `Grid` that are in range [`Min`,`Max`].
- `CopyProperty GridSource::PropSource::GridTarget::`
  `PropTarget[::Overwrite::isHardData]` Copy a property from one object (point-set or Cartesian grid) to another. `PropSource` is the property to be copied, from object `GridSource` to object `GridTarget`. The copied property is called `PropTarget`. If optional parameter `Overwrite` is equal to 1, the copy would overwrite values already on `PropTarget`. Parameter `isHardData` sets the copied values as hard data if `isHardData=1`. By default, `Overwrite=0` and `isHardData=0`.
- `DeleteObjectProperties Grid::Prop1[::Prop2::....::PropN]`
  Delete all the specified properties from `Grid`.

- `DeleteObjects Grid1[::Grid2::...::GridN]` Delete the specified objects.
- `SwapPropertyToRAM Grid::Prop1[::Prop2::...::PropN]` Loads the specified properties into the random access memory (RAM). Operating on properties loaded in RAM is faster than working with properties accessed from the disk.
- `SwapPropertyToDisk Grid::Prop1[::Prop2::...::PropN]` Removes the specified properties from RAM and stores them on the disk. Accessing those properties, for example displaying them, will be slower. This is useful to control RAM consumption. The SGeMS simulation algorithms (SGSIM, SISIM, etc.) store all but the latest realization on the disk.
- `LoadProject ProjectName` Load the specified project. The `ProjectName` must be given as a full path folder, e.g. `D:/user/test2d.prj/`.
- `SaveGeostatGrid Grid::Filename::Filter` Save the specified grid on file. The `Filename` must be given with a full directory path. `Filter` specifies the data format: either ASCII GSLIB or binary *sgems*.
- `LoadObjectFromFile Filename` Load the object from the specified file. The `Filename` must be given with a full directory path.
- `NewCartesianGrid Name::Nx::Ny::Nz[::SizeX::SizeY::SizeZ]` `[::Ox::Oy::Oz]` Create a new Cartesian grid with the specified geometry. The default pixel size value for [`SizeX`, `SizeY`, `SizeZ`] is 1 and the default origin [`Ox`, `Oy`, `Oz`] is 0.
- `RotateCamera x::y::z::angle` Rotate the camera: (x,y,z) defines the rotation axis; the rotation `angle` is in radians and measured clockwise.
- `SaveCameraSettings Filename` Save the position of the camera into `Filename`. The `Filename` must be given with a full directory path.
- `LoadCameraSettings Filename` Retrieve the camera position from `Filename`. The `Filename` must be given with a full directory path.
- `ResizeCameraWindow Width::Height` Set the width and the height of the camera to `Width` and `Height`.
- `ShowHistogram Grid::Prop[::NumberBins::LogScale]` Display the histogram of the specified property. The number of bins may also be input with `NumberBins`; the default value is 20. The x axis can also be changed to log scale by setting `LogScale` to true (value 1).
- `SaveHistogram Grid::Prop::Filename[::Format]` `[::NumberBins][::LogScale][::ShowStats][::ShowGrid]` Save the specified histogram into `Filename` with format specified by `Format`. The available file formats are PNG and BMP. The default format is PNG. When `ShowStats` is set to true (value 1), it saves the statistics in the file, `ShowGrid` adds a grid to the histogram. The `Filename` must be given with a full directory path.
- `SaveQQplot Grid::Prop1::Prop2::Filename[::Format]` `[::ShowStats][::ShowGrid]` Save the QQ plot between `Prop1` and `Prop2` into

`Filename`. The available file formats are PNG and BMP. The default format is PNG. When `ShowStats` is set to true (value 1), it saves the statistics in the file, `ShowGrid` adds a grid to the QQ plot. The `Filename` must be given with a full directory path.

- `SaveScatterplot Grid::Prop1::Prop2::Filename`
  `[::Format][::ShowStats][::ShowGrid][::YLogScale:: XLogScale]` Save the scatter plot between `Prop1` and `Prop2` into `Filename`. The available file formats are PNG and BMP. The default format is PNG. When `ShowStats` is set to true (value 1), it saves the statistics in the file, `ShowGrid` adds a grid to the scatter plot. The `Filename` must be given with a full directory path.

- `DisplayObject Grid[::Prop]` Display `Prop` on the viewing window. When `Prop` is not specified, only the grid geometry is displayed.

- `HideObject Grid` Remove `Grid` from the viewing window.

- `TakeSnapshot Filename[::Format]` Take the snapshot of the current viewing window and save it into `Filename`. The available file formats are PNG, PS, BMP and PPM. The default format is PNG. The `Filename` must be given with a full directory path.

- `RunGeostatAlgorithm AlgorithmName::ParametersHandler::`
  `AlgorithmParameters` Run the algorithm specified by `AlgorithmName`. `ParametersHandler` tells SGeMS how to parse the parameters. Use `/GeostatParamUtils/XML` unless you know what you are doing. The parameters of the algorithm are provided as an XML string, identical to the one obtained by saving the algorithm parameters to a file (the only difference is that all the parameters must be on a single line).

10.1.2 Execute command file

As mentioned previously it is possible to create a commands file which contains a list of commands to be performed. A commands file is executed by clicking the *Execute Commands File...* button and browsing to the file location. SGeMS will automatically run all the commands, in the order they appear. Each command in the script file must start on a new line and be contained on a **single** line; lines that start with a # sign are ignored. Figure 10.2 shows a simple example that loads a SGeMS project, creates a Cartesian grid, runs the *SGSIM* algorithm over that grid, and finally saves the grid with all its properties in *gslib* format. Note that for formatting purposes, the parameters of the `RunGeostatAlgorithm` command span multiple lines. In an actual command file, all these parameters should be on a single line.

```
#Load a SGeMS project, run sgsim and save data in gslib format
LoadObjectFromFile D:/program/test.prj/TI::s-gems NewCartesianGrid
grid::250::250::1::1.0::1.0::1.0::0::0::0 RunGeostatAlgorithm
sgsim::/GeostatParamUtils/XML::<parameters>
    <algorithm name="sgsim"/>
    <Grid_Name value="grid"/>
    <Property_Name value="sgsim"/>
    <Nb_Realizations value="1"/>
    <Seed value="14071789"/>
    <Kriging_Type value="Simple Kriging (SK)"/>
    <Trend value="0 0 0 0 0 0 0 0 0"/> <Local_Mean_Property value=""/>
    <Assign_Hard_Data value="1"/> <Hard_Data grid="" property=""/>
    <Max_Conditioning_Data value="20"/>
    <Search_Ellipsoid value="40 40 1 0 0 0"/>
    <Use_Target_Histogram value="0"/>
    <nonParamCdf ref_on_file ="0" ref_on_grid ="1"
        filename ="" grid ="" property =""> <LTI_type function ="Power"
        extreme ="0" omega ="3"/> <UTI_type function ="Power" extreme
        ="0" omega ="0.333"/>
    </nonParamCdf>
    <Variogram nugget="0.1" structures_count="1">
        <structure_1 contribution="0.9" type="Spherical">
            <ranges max="30" medium="30" min="1"/>
            <angles x="0" y="0" z="0"/>
        </structure_1>
    </Variogram>
  </parameters>
SaveGeostatGrid grid::D:/program/sgsim.out::gslib
```

Figure 10.2 An example of command script file

Commands files are an easy way to automate tasks in SGeMS. They, however, offer limited flexibility: there is indeed no support for control structures such as loops or tests. Hence performing twenty runs of a given algorithm, each time changing a parameter, would require that the twenty corresponding commands be written in the commands file. To address the need for such more powerful scripting capabilities, SGeMS supports another automation mechanism: Python scripts.

10.2 Python script

Python is a popular scripting language which provides all the facilities expected from modern programming languages such as variable definition, function definition or object-oriented programming. Please refer to www.python.org for resources on Python, including tutorials and extension libraries. By embedding a

Python interpreter, SGeMS provides a powerful way of performing repetitive tasks and even extending its functionalities.

10.2.1 SGeMS Python modules

Python can interact with SGeMS through the three functions defined by the `sgems` module.

- `execute( 'Command' )` Executes SGeMS command `Command`. This achieves the same result as typing command `Command` in the command line editor (see Section 10.1). This function is very useful as it enables you to programmatically control SGeMS: it is possible to run several geostatistics algorithms with dynamically set parameters, display the results, save a screen capture, plot histograms, etc.
- `get_property( GridName, PropertyName )` Returns an array (actually a *list* in Python terminology) containing the values of property `PropertyName` of object `GridName`.
- `set_property( GridName, PropertyName, Data )` Sets the values of property `PropertyName` of object `GridName` to the values of *list* `Data`. If object `GridName` has no property called `PropertyName`, a new property is created.

10.2.2 Running Python scripts

Python script files are executed by selecting *Run Script...* from the *Scripts* menu and browsing to the script location. SGeMS also provides a rudimentary script editor that can directly execute scripts by pressing its *Run* button or by pressing the *F5* key. Outputs generated by the script, e.g. messages, warnings, errors, are displayed in the *Scripts Output* tab of the Commands Panel, and, if the scripts editor is used, in the *Script Output Messages* section of the editor.

Figure 10.3 is an example script that computes the logarithm of values taken from property `samples` of a grid named `grid`. It then writes the logarithms to a new property called `log_samples`, displays that property in the 3D view and saves a snapshot in PNG format.

Providing an introduction to Python is beyond the scope of this chapter, and the reader is encouraged to read the tutorial available at `http://docs.python.org/tut/tut.html`. However, the script of Fig. 10.3 is simple enough so hopefully a beginner can understand the following line-by-line explanation.

line 1 Python scripts that interact with SGeMS should always start with this line: it makes the SGeMS-specific functions (see Section 10.2.1) available to Python.

```
1   import sgems
2   from math import *
3
4   data = sgems.get_property('grid', 'samples')
5
6    for i in range( len(data) ):
7      if data[i]>0 :
8        data[i] = log(data[i])
9      else:
10       data[i] = -9966699
11
12  sgems.set_property('grid', 'log_samples', data )
13  sgems.execute( 'DisplayObject grid::log_samples' )
14  sgems.execute( 'TakeSnapshot log_samples.png::PNG' )
```

Figure 10.3 Python script example

line 2 load all the functions defined in the standard `math` library (module).

line 4 copy the values of property `samples` of grid `grid` into list `data`.

lines 6–10 iterate through all the values in `data`. If the current value `data[i]` is strictly positive, take its logarithm, otherwise, set the value to the default code for no-data-value: -9966699. List `data` now contains the logarithms of the `samples` property.

line 12 write the log values to a new property of grid `grid`, called `log_samples`. Function `set_property` is one of the functions defined by the `sgems` module. Note that if object `grid` already had a property called `log_samples`, it would have been over-written.

line 13 use the `execute` function defined by module `sgems` to run SGeMS command `DisplayObject` (see Section 10.1.1 for a description of the SGeMS commands).

line 14 use the `execute` function to run SGeMS command `TakeSnapshot`.

Figure 10.4 shows another Python script used to calculate the calibration coefficient required by the Markov–Bayes model in *COSISIM* (Section 8.1.6). The original Ely point-set data, both primary and secondary (Section 4.1.1), have been transformed into indicator variables giving thresholds in line 4 (Fig. 10.4).

Many libraries are available to extend the capabilities of Python, for example to execute system commands (package *os*), display graphs (package *matplotlib*, http://matplotlib.sourceforge.net), build simple graphical interfaces (package *wxPython*, http://www.wxpython.org) or compute a Fast Fourier transform (package *SciPy*, http://www.scipy.org).

```
1  import sgems # import sgems modulus
2
3  # thresholds $z_k$s
4  tr = [3.5, 4, 4.5, 5, 5.5, 6, 6.5, 7, 7.5, 8]
5
6  for t in tr: # for each threshold
7    m0 = 0 # $m^0(Z)$
8    m1 = 0 # $m^1(Z)$
9    cnt_p = 0 # nb of data when $I(u;z_k)$=1
10
11   # get the secondary data
12   sec = sgems.get_property('Ely1_pset','Secondary_soft_'+str(t))
13   # get the primary data
14   prim = sgems.get_property('Ely1_pset','Primary_id'+str(t))
15
16   for p,s, in zip(prim,sec): # for each data pair
17     cnt_p += p # p value is either 1 or 0
18     if p == 1: # when I(u; z)=1
19       m1 += s # add secondary data vale
20     else: # when I(u; z)=0
21       m0 += s
22
23   # output B(z) value
24   print m1/cnt_p - m0/(len(prim)-cnt_p)
```

Figure 10.4 Python script to calculate the calibration coefficient values for Markov–Bayes model in *COSISIM*

Notice that SGeMS provides a minimal install of Python. If you wish to install Python and extension libraries, be sure to delete the Python dynamic library (called python.dll on Windows systems and installed in the same directory as the sgems executable) included in the SGeMS distribution.

10.3 Plug-ins

One of the objectives of SGeMS is to provide a flexible platform in which new ideas can be conveniently implemented and made available through a user-friendly interface. Integrating the Python scripting language into SGeMS is a first answer to that requirement. Python enables quick prototyping of algorithms that capitalize on the existing tools of SGeMS, for example a history matching algorithm that iteratively generates stochastic realizations of a permeability field based on the output of a flow simulator. However, Python scripts have two main limitations: scripts can not capitalize on the SGeMS graphical interface to gather user input parameters, and Python does not have access to the low-level capabilities of SGeMS. For example, the many algorithms coded in SGeMS for searching neighboring data on a grid (Cartesian grid or set of points) are not accessible to Python scripts.

These limitations are alleviated by using the SGeMS *plug-in* mechanism. A plug-in is a piece of software that does not run by itself but is automatically recognized and executed by SGeMS. Plug-ins can be used to add new geostatistics algorithms (the new algorithms appear in the algorithms list, just like the standard SGeMS geostatistics tools) or new input/output filters to support more data file formats. Plug-ins can even extend the set of supported grid types beyond Cartesian grids and sets of points. Unlike Python scripts, plug-ins can use the entire SGeMS API (i.e. all the C++ objects and methods defined in SGeMS), enabling faster and more robust implementation of new algorithms: tested routines for common operations such as neighbors retrieval or Monte Carlo sampling from a distribution are readily available and do not have to be implemented anew.

Since SGeMS aims to be a development platform for new geostatistics ideas, plug-ins defining new algorithms are of particular interest. These are usually composed of two files: a dynamic library file (often called *dll* in Microsoft Windows) which contains all the compiled code for the algorithm, and a text file with extension `.ui` that describes a graphical interface (for user parameters input). An algorithm plug-in is installed by placing these two files in the `plugins/Geostat` directory of SGeMS: on Microsoft Windows, if SGeMS was installed in `C:/Program Files/SGeMS`, the files should be placed in `C:/Program Files/SGeMS/plugins/Geostat`. When SGeMS is started, it automatically recognizes the new plug-in.

It is beyond the scope of this book to explain how to write plug-ins. The interested reader is invited to refer to the SGeMS website (`sgems.sourceforge.net`) for tutorials and references.

Plug-ins are the favored solution to develop and distribute new algorithms within SGeMS.

Bibliography

Alabert, F. G. 1987, The practice of fast conditional simulations through the LU decomposition of the covariance-matrix, *Mathematical Geology* **19**(5), 369–386.

Almeida, A. S. and Journel, A. G. 1994, Joint simulation of multiple variables with a Markov-type coregionalization model, *Mathematical Geology* **26**(5), 565–588.

Anderson, T. W. 2003, *An Introduction to Multivariate Statistical Analysis*, 3rd edn, New York, John Wiley & Sons.

Armstrong, M., Galli, A. G., Le Loc'h, G., Geffroy, F. and Eschard, R. 2003, *Plurigaussian Simulations in Geosciences*, Berlin, Springer.

Arpat, B. 2004, *Sequential simulation with patterns*, Ph.D. thesis, Stanford University, Stanford, CA.

Barnes, R. and Johnson, T. 1984, Positive kriging, in G. Verly *et al.* (eds.), *Geostatistics for Natural Resources Characterization*, Vol. 1, Dordrecht, Holland, Reidel, pp. 231–244.

Benediktsson, J. A. and Swain, P. H. 1992, Consensus theoretic classification methods, *IEEE Transactions on System, Man and Cybernetics* **22**(4), 688–704.

Bordley, R. F. 1982, A multiplicative formula for aggregating probability assessments, *Management Science* **28**(10), 1137–1148.

Boucher, A. and Kyriakidis, P. C. 2006, Super-resolution land cover mapping with indicator geostatistics, *Remote Sensing of Environment* **104**(3), 264–282.

Bourgault, G. 1994, Robustness of noise filtering by kriging analysis, *Mathematical Geology* **26**, 733–752.

Bourgault, G. 1997, Using non-Gaussian distributions in geostatistical simulations, *Mathematical Geology* **29**(3), 315–334.

Caers, J. 2005, *Petroleum Geostatistics*, Society of Petroleum Engineers.

Caers, J. and Hoffman, T. 2006, The probability perturbation method: a new look at bayesian inverse modeling, *Mathematical Geology* **38**(1), 81–100.

Canny, J. 1986, A computational approach to edge detection, *IEEE Transactions on Pattern Analysis and Machine Intelligence* **8**(6), 679–698.

Castro, S. 2007, *A Probabilistic Approach to Jointly Integrate 3D/4D Seismic, Production Data and Geological Information for Building Reservoir Models*, Ph.D. thesis, Stanford University, Stanford, CA.

Chauvet, P. 1982, The variogram cloud, in T. Johnson and R. Barnes (eds.), *Proceedings of the 17th APCOM International Symposium, Golden, Colorado*, Society of Mining Engineers, New York, pp. 757–764.

254

Chilès, J. and Delfiner, P. 1999, *Geostatistics: Modeling Spatial Uncertainty*, New York, John Wiley & Sons.

Christakos, G. 1984, On the problem of permissible covariance and variogrammodels, *Water Resources Research* **20**(2), 251–265.

Cressie, N. 1993, *Statistics for Spatial Data*, New York, John Wiley & Sons.

Daly, C. and Verly, G. W. 1994, Geostatistics and data integration, in R. Dimitrakopoulos (ed.), *Geostatistics for the Next Century*, Kluwer, pp. 94–107.

David, M. 1977, *Geostatistical Ore Reserve Estimation*, Amsterdam, Elsevier.

Davis, M. 1987, Production of conditional simulations via the LU decomposition of the covariance matrix, *Mathematical Geology* **19**(2), 91–98.

Deutsch, C. V. 1994, Algorithmically-defined random function models, in Dimitrakopoulos (ed.), *Geostatistics for the Next Century*, Dordrecht, Holland, Kluwer, pp. 422–435.

Deutsch, C. V. 1996, Constrained modeling of histograms and cross plots with simulated annealing, *Technometrics* **38**(3), 266–274.

Deutsch, C. V. 2002, *Geostatistical Reservoir Modeling*, New York, Oxford University Press.

Deutsch, C. V. and Journel, A. G. 1998, *GSLIB: Geostatistical Software Library and User's Guide*, 2nd edn, New York, Oxford University Press.

Deutsch, C. V. and Tran, T. T. 2002, Fluvsim: a program for object-based stochastic modeling of fluvial depositional system, *Computers & Geosciences* **28**(4), 525–535.

Dietrich, C. R. and Newsam, G. N. 1993, A fast and exact method for multidimensional Gaussian stochastic simulation, *Water Resource Research* **29**(8), 2861–2869.

Dimitrakopoulos, R. and Luo, X. 2004, Generalized sequential Gaussian simulation, *Mathematical Geology* **36**, 567–591.

Dubrule, O. 1994, Estimating or choosing a geostatistical model?, in R. Dimitrakopoulos (ed.), *Geostatistics for the Next Century*, Kluwer, pp. 3–14.

Farmer, C. 1992, Numerical rocks, in P. King (ed.), *The Mathematical Generation of Reservoir Geology*, Oxford, Clarendon Press.

Frigo, M. and Johnson, S. G. 2005, The design and implementation of FFTW3, *Proceedings of the IEEE* **93**(2), 216–231.

Gloaguen, E., Marcotte, D., Chouteau, M. and Perroud, H. 2005, Borehole radar velocity inversion using cokriging and cosimulation, *Journal of Applied Geophysics* **57**, 242–259.

Goldberger, A. 1962, Best linear unbiased prediction in the generalized linear regression model, *Journal of the American Statistical Association* **57**, 369–375.

Gómez-Hernández, J. J. and Cassiraga, E. F. 1994, Theory and practice of sequential simulation, in M. Armstrong and P. Dowd (eds.), *Geostatistical Simulations*, Dordrecht, Holland, Kluwer Academic Publishers, pp. 111–124.

Gómez-Hernández, J. J. and Journel, A. G. 1993, Joint sequential simulation of multiGaussian fields, in A. Soares (ed.), *Geostatistics-Troia*, Vol. 1, Dordrecht, Kluwer Academic Publishers, pp. 85–94.

Gómez-Hernández, J. J., Froidevaux, R. and Biver, P. 2005, Exact conditioning to linear constraints in kriging and simulation, in O. Leuangthong and C. V. Deutsch (eds.), *Geostatistics Banff 2004*, Vol. 2, Springer, pp. 999–1005.

Goovaerts, P. 1997, *Geostatistics for Natural Resources Evaluation*, New York, Oxford University Press.

Guardiano, F. and Srivastava, R. M. 1993, Multivariate geostatistics: beyond bivariate moments, in A. Soares (ed.), *Geostatistics-Troia*, Vol. 1, Dordrecht, Kluwer Academic Publishers, pp. 133–144.

Haldorsen, H. H. and Damsleth, E. 1990, Stochastic modeling, *Journal of Petroleum Technology*, pp. 404–412.

Hansen, T. M., Journel, A. G., Tarantola, A. and Mosegaard, K. 2006, Linear inverse Gaussian theory and geostatistics, *Geophysics* **71**(6), R101–R111.

Hartigan, J. A. 1975, *Clustering Algorithms*, New York, John Wiley & Sons Inc.

Hu, L. Y., Blanc, G. and Noetinger, B. 2001, Gradual deformation and iterative calibration of sequential stochastic simulations, *Mathematical Geology* **33**(4), 475–489.

Isaaks, E. H. 2005, The kriging oxymoron: a conditionally unbiased and accurate predictor (2nd edn), in O. Leuangthong and C. V. Deutsch (eds.), *Geostatistics Banff 2004*, Vol. 1, Springer, pp. 363–374.

Isaaks, E. H. and Srivastava, R. M. 1989, *An Introduction to Applied Geostatistics*, New York, Oxford University Press.

Jensen, J. L., Lake, L. W., Patrick, W. C. and Goggin, D. J. 1997, *Statistics for Petroleum Engineers and Geoscientists*, New Jersey, Prentice Hall.

Johnson, M. 1987, *Multivariate Statistical Simulation*, New York, John Wiley & Sons.

Jolliffe, I. T. 1986, *Principal Component Analysis*, New York, Springer-Verlag.

Journel, A. G. 1980, The lognormal approach to predicting local distribution of selective mining unit grades, *Mathematical Geology* **12**(4), 285–303.

Journel, A. G. 1983, Non-parametric estimation of spatial distributions, *Mathematical Geology* **15**(3), 793–806.

Journel, A. G. 1986, Geostatistics: models and tools for the earth sciences, *Mathematical Geology* **18**(1), 119–140.

Journel, A. G. 1989, *Fundamentals of Geostatistics in Five Lessons*, Vol. 8, Short Course in Geology, Washington, D.C., American Geophysical Union.

Journel, A. G. 1993, Geostatistics: roadblocks and challenges, in A. Soares (ed.), *Geostatistics-Troia*, Vol. 1, Dordrecht, Kluwer Academic Publishers, pp. 213–224.

Journel, A. G. 1994, Modeling uncertainty: some conceptual thoughts, in R. Dimitrakopoulos (ed.), *Geostatistics for the Next Century*, Kluwer, pp. 30–43.

Journel, A. G. 1999, Markov models for cross covariances, *Mathematical Geology* **31**(8), 955–964.

Journel, A. G. 2002, Combining knowledge from diverse sources: An alternative to traditional data independence hypotheses, *Mathematical Geology* **34**(5), 573–596.

Journel, A. G. and Alabert, F. G. 1989, Non-Gaussian data expansion in the earth sciences, *Terra Nova* **1**, 123–134.

Journel, A. G. and Deutsch, C. 1993, Entropy and spatial disorder, *Mathematical Geology* **25**(3), 329–355.

Journel, A. G. and Froidevaux, R. 1982, Anisotropic hole-effect modeling, *Mathematical Geology* **14**(3), 217–239.

Journel, A. G. and Huijbregts, C. J. 1978, *Mining Geostatistics*, New York, Academic Press.

Journel, A. G. and Kyriakidis, P. C. 2004, *Evaluation of Mineral Reserves: A Simulation Approach*, New York, Oxford University Press.

Journel, A. G. and Rossi, M. E. 1989, When do we need a trend model?, *Mathematical Geology* **21**(7), 715–739.

Journel, A. G. and Xu, W. 1994, Posterior identification of histograms conditional to local data, *Mathematical Geology* **26**, 323–359.

Journel, A. G. and Zhang, T. 2006, The necessity of a multiple-point prior model, *Mathematical Geology* **38**(5), 591–610.

Koch, G. S. and Link, R. F. 1970, *Statistical Analysis of Geological Data*, John Wiley and Sons Inc.

Krige, D. G. 1951, *A statistical approach to some mine valuations and allied problems at the Witwatersrand*, M.S. thesis, University of Witwatersrand, South Africa.

Krishnan, S. 2004, *Combining diverse and partially redundant information in the earth sciences*, Ph.D. thesis, Stanford University, Stanford, CA.

Krishnan, S., Boucher, A. and Journel, A. G. 2005, Evaluating information redundancy through the tau model, in O. Leuangthong and C. V. Deutsch (eds.), *Geostatistics Banff 2004*, Vol. 2, Springer, pp. 1037–1046.

Kyriakidis, P. C. and Yoo, E. H. 2005, Geostatistical prediction and simulation of point values from areal data, *Geographical Analysis* **37**(2), 124–151.

Kyriakidis, P. C., Schneider, P. and Goodchild, M. F. 2005, Fast geostatistical areal interpolation, *7th International Conference on Geocomputation*, Ann Arbor, Michigan.

Lantuéjoul, C. 2002, *Geostatistical Simulation: Models and Algorithms*, Berlin, Germany, Springer-Verlag.

Liu, Y. 2007, *Geostatistical Integration of Coarse-scale Data and Fine-scale Data*, Ph.D. thesis, Stanford University, Stanford, CA.

Liu, Y. and Journel, A. G. 2005, Average data integration (implementation and case study), *Report 18 of Stanford Center for Reservoir Forecasting*, Stanford, CA.

Liu, Y., Jiang, Y. and Kyriakidis, P. 2006a, Calculation of average covariance using Fast Fourier Transform (fft), *Report 19 of Stanford Center for Reservoir Forecasting*, Stanford, CA.

Liu, Y., Journel, A. G. and Mukerji, T. 2006b, Geostatistical cosimulation and downscaling conditioned to block data: Application to integrating vsp, travel-time tomography, and well data, *SEG Technical Program Expanded Abstracts*, pp. 3320–3324.

Luenberger, D. G. 1969, *Optimization by Vector Space Methods*, New York, John Wiley & Sons.

Maharaja, A. 2004, *Hierarchical simulation of multiple facies reservoir using multiple-point geostatistics*, M.S. thesis, Stanford University, Stanford, CA.

Mallet, J. L. 2002, *Geomodeling*, New York, Oxford University Press.

Matheron, G. 1962, Traité de géostatistique appliquée, tome ii. Vol. 1, ed. Technip, Paris.

Matheron, G. 1963, Traité de géostatistique appliquée, tome ii. Vol. 2, ed. Technip, Paris.

Matheron, G. 1970, La théorie des variables régionalisées, et ses applications. Les cahiers du Centre de Morphologie Mathématique de Fontainebleau, Fascicule 5.

Matheron, G. 1973, The intrinsic random functions and their applications, *Advances in Applied Probability* **5**, 439–468.

Matheron, G. 1978, Estimer et choisir, *Technical report*, Fascicules n7, Les cahiers du Centre de Morphologie Mathématique de Fontainebleau, Ecole des Mines de Paris.

Myers, D. E. 1982, Matrix formulation of co-kriging, *Mathematical Geology* **14**(3), 249–257.

Nowak, W., Tenkleve, S. and Cirpka, O. A. 2003, Efficient computation of linearized cross-covariance and auto-covariance matrices of interdependent quantities, *Mathematical Geology* **35**, 53–66.

Olea, R. A. 1999, *Geostatistics for Engineers and Earth Scientists*, Kluwer Academic Publishers.

Oliver, D. S. 1995, Moving averages for Gaussian simulation in two and three dimensions, *Mathematical Geology* **27**(8), 939–960.

Oz, B., Deutsch, C. V., Tran, T. T. and Xie, Y. L. 2003, DSSIM-HR: A FORTRAN 90 program for direct sequential simulation with histogram reproduction, *Computers & Geosciences* **29**(1), 39–51.

Polyakova, E. and Journel, A. G. in press, The nu expression for probabilistic data integration, *Mathematical Geology*.

Rao, S. and Journel, A. G. 1996, Deriving conditional distributions from ordinary kriging, in E. Baffi and N. Shofield (eds.), *Fifth International Geostatistics Congress*, Wollongong, Kluwer Academic Publishers.

Remy, N. 2001, Post-processing a dirty image using a training image, *Report 14 of Stanford Center for Reservoir Forecasting*, Stanford University, Stanford, CA.

Rendu, J.-M. M. 1979, Normal and lognormal estimation, *Mathematical Geology* **11**(4), 407–422.

Rivoirard, J. 2004, On some simplifications of cokriging neighborhood, *Mathematical Geology* **36**(8), 899–915.

Rosenblatt, M. 1952, Remarks on a multivariate transformation, *Annals of Mathematical Statistics* **23**(3), 470–472.

Schneiderman, H. and Kanade, T. 2004, Object detection using the statistics of parts, *International Journal of Computer Vision* **56**(3), 151–177.

Schnetzler, E. 1994, *Visualization and cleaning of pixel-based images*, M.S. thesis, Stanford University, Stanford, CA.

Soares, A. 2001, Direct sequential simulation and cosimulation, *Mathematical Geology* **33**(8), 911–926.

Srivastava, R. M. 1987, Minimum variance or maximum profitability? *CIM Bulletin* **80**(901), 63–68.

Srivastava, R. M. 1994, The visualization of spatial uncertainty, in J. Yarus and R. Chambers (eds.), *Stochastic Modeling and Geostatistics: Principles, Methods and Case Studies*, Vol. 3, AAPG, pp. 339–345.

Stoyan, D., Kendall, W. S. and Mecke, J. 1987, *Stochastic Geometry and its Applications*, New York, John Wiley & Sons.

Strebelle, S. 2000, *Sequential simulation drawing structures from training images*, Ph.D. thesis, Stanford University, Stanford, CA.

Strebelle, S. 2002, Conditional simulation of complex geological structures using multiple-point statistics, *Mathematical Geology* **34**(1), 1–21.

Tarantola, A. 2005, *Inverse Problem Theory and Methods for Model Parameter Estimation*, Philadelphia, Society for Industrial and Applied Mathematics.

Tran, T. T. 1994, Improving variogram reproduction on dense simulation grids, *Computers & Geosciences* **20**(7), 1161–1168.

Vargas-Guzman, J. A. and Dimitrakopoulos, R. 2003, Computational properties of min/max autocorrelation factors, *Computers & Geosciences* **29**(6), 715–723.

Wackernagel, H. 1995, *Multivariate Statistics*, Berlin, Springer-Verlag.

Walker, R. 1984, General introduction: Facies, facies sequences, and facies models, in R. Walker (ed.), *Facies Models*, 2nd edn, Geoscience Canada Reprint Series 1, Toronto, Geological Association of Canada, pp. 1–9.

Wu, J., Boucher, A. and Zhang, T. in press, A sgems code for pattern simulation of continuous and categorical variables: Filtersim, *Computers & Geosciences*.

Yao, T. T. and Journel, A. G. 1998, Automatic modeling of (cross) covariance tables using fast Fourier transform, *Mathematical Geology* **30**(6), 589–615.

Zhang, T. 2006, *Filter-based Training Pattern Classification for Spatial Pattern Simulation*, Ph.D. thesis, Stanford University, Stanford, CA.

Zhang, T., Journel, A. G. and Switzer, P. 2006, Filter-based classification of training image patterns for spatial simulation, *Mathematical Geology* **38**(1), 63–80.

Zhu, H. and Journel, A. G. 1993, Formatting and interpreting soft data: stochastic imaging via the Markov–Bayes algorithm, in A. Soares (ed.), *Geostatistics-Troia*, Vol. 1, Dordrecht, Kluwer Academic Publishers, pp. 1–12.

Index